T0140162

Guide to Deep Learning Basics

Sandro Skansi
Editor

Guide to Deep Learning Basics

Logical, Historical and Philosophical Perspectives

Editor
Sandro Skansi
Faculty of Croatian Studies
University of Zagreb
Zagreb, Croatia

ISBN 978-3-030-37593-5 ISBN 978-3-030-37591-1 (eBook)
https://doi.org/10.1007/978-3-030-37591-1

This Springer imprint is published by the registered company Springer Nature Switzerland AG
The registered company address is: Gewerbestrasse 11, 6330 Cham, Switzerland

Preface

Artificial neural networks came to existence in 1943 with the seminal paper by Walter Pitts and Warren McCulloch. It is commonly said that the rest is history. But this history, which is seldom explored, holds many interesting details. From the purely historical unknowns to the conceptual connections often spanning back to medieval and even classical times which we often take for granted. By doing so, we often simplify things to a degree when it is no longer evident how rich and intricate the history (and prehistory) of deep learning was. The present volume brings new light on some foundational issues, and we hope that it will shed a new light on this amazing field of research.

My personal pivotal point for editing this volume was the discovery of a lost Croatian machine translation project from 1959. It was interesting to see how I was rediscovering the ideas that were so geographically close, and yet so remote and lost. But one question arose: If there was a whole project in machine translation no one knows about, what else is there to dig out? Can we find new old ideas that contribute to the rich history deep learning? Or more generally, how did this amazing field survive against the tide and finally flourish? Such complex history is bound to have nooks and crannies just waiting to be rediscovered and explored like a lost city of a bygone civilization.

But we could ask this general historical and "archeological" question in a different way, making it sound more philosophical and analytical: Was the success of deep learning wholly due to its technological superiority? Or was deep learning conceptually a better theory, but it did not prosper due to high computational needs not available back in the day? Following McCarthy, GOFAI separated itself from its philosophical backgrounds, and by doing so it shed layer after layer of what was seen as "computational inefficiencies". But was this (past) methodological necessity also a conceptual necessity? Deep learning more often than not embraced its "humanistic call" and tried hard not to dismiss problems that were too vague or imprecise to tackle, even at the cost of not producing working systems. It can be argued, as some authors of this monograph do, that this focus on concepts rather than working production systems has paved the way for the dramatic rise of deep learning, and in turn enabled it to surpass GOFAI and develop better working AI systems.

One could ask what could such a book do for a technical field such as deep learning. It is my deep belief that as technology progresses, it needs philosophy and even art to make room for it in the common culture and in the everyday lives of people. And this is a challenge for the professional, not for the common person. People need to feel the need for science and technology, and welcome it in their lives without fear or reservation. And this can only be done if the scientist and coder are able to take a step back from its lab, GPU's and code, and explain the "why". It is our hope that this book will help in doing so.

This book is not technical, and definitely not made to be inaccessible, but since it explores very specific and sometimes demanding ideas from a very abstract perspective it might come across daunting. Some people are geared in such a way that they like this perspective and need to see the big picture first before digging in into the technical details. Other people will not take kindly on philosophizing without first tackling and idea heads-on and coding. I can guarantee that both of these types of people will find interesting topics explored here, but the ideal reader of this volume is a wholly different person. She is someone who knows deep learning, and has spent countless hours studying it and coding. By knowing how deep learning systems are made and exactly how they work, she has become somewhat disenchanted by it. She is thinking that deep learning and AI in general is actually simple computation. It is our hope, by going sometimes far beyond actual applications, that we will rekindle her passion for this unique discipline and show that even when everything technical is mastered, and the magic goes away, that there is still something special, unique, and mystical on the very edges of deep learning. The ideas presented here are lights and echoes over the horizon.

As a final note, I would like to note that the authors of the chapters who are affiliated with the University of Zagreb but are not current employees of the university have their affiliation written as "University of Zagreb" with no faculty or department. I would also like to thank the reviewers and editors at Springer for all their help. A volume such as this is necessarily incomplete (since there are many interesting facets not covered), and I hold only myself accountable for any such incompleteness.

Samobor, Croatia Sandro Skansi
October 2019

Contents

Chapter 1
Mathematical Logic: Mathematics of Logic or Logic of Mathematics

Zvonimir Šikić

Abstract This brief historical survey is written from a logical point of view. It is a rational reconstruction of the genesis of some interrelations between formal logic and mathematics. We examine how mathematical logic was conceived: as the abstract mathematics of logic or as the logic of mathematical practice.

Keywords Mathematical logic · Logic of mathematics · Mathematics of logic · Formalizations · Laws of thought

At the end of the first half of nineteenth century formal logic was being developed mainly in Great Britain, under the influence of the newly established abstract algebra. It has just begun to appear in the works of Peacock [15], De Morgan [3], and Hamilton [12]. Attempts to apply mathematical analysis (which proved fruitful and convincing in studying the laws of quantities) to formal logic (which deals with the laws of qualities) became characteristic for this period. Setting a suitable symbolic apparatus and founding the laws of its manipulation, in the same way arithmetic does, was the ultimate purpose of the use of the new method. Similar attempts can be found in earlier periods and other places (e.g., Leibniz [14] on the continent) but the first great success was achieved in 1847 by Boole in [1]. It was Boole who showed that the fundamental operations with concepts (operations of thought, he would say) can be represented by the arithmetic operations of addition, subtraction, and multiplication; while the fundamental concepts of "everything" and "nothing" can be represented by 1 and 0. These operations are governed by the arithmetic laws, i.e., by the laws of quantities, together with the additional law $x^2 = x$, called the principle of tautology. This principle is characteristic and distinctive for operations with qualities. Thus the logic of concepts (or classes) became special arithmetic and it is in this way that Boole understands it. Premises are equations which, using arithmetic operations, can produce other equations, the latter being the conclusions of this arithmetical reasoning. The problem of the logical interpretation of the equations is solved by reducing the equations to its normal form, by applying Taylor's formula characteristic for this $(x^2 = x)$–arithmetic:

Z. Šikić (✉)
University of Zagreb, Zagreb, Croatia
e-mail: zvonimir.sikic@gmail.com

S. Skansi (ed.), *Guide to Deep Learning Basics*,
https://doi.org/10.1007/978-3-030-37591-1_1

$$f(x) = f(1) \cdot x + f(0) \cdot (1 - x)$$

By this analysis, the logic of concepts acquired simplicity, safety, and generality of arithmetic. In other words, it was well, and that meant arithmetically, founded. Equations which was not possible to interpret in the course of the proof were a disturbance to the purely logical understanding of the basic laws. Hence, there began dearithmetization of formal logic, by Boole himself [2], Jevons [13], Peirce [21], Venn [29], and Schröder [26] (the title of Jevons' [13] is noteworthy). The final result was the Boole–Schröder algebra which represented, in its various interpretations, the logic of concepts, the logic of one-place propositional functions and the logic of propositions (starting with C. S. Peirce and E. Schröder).

Dearithmetization of formal logic has not changed the fundamental character of its foundation (which is often thought and said on the basis of the later logistic or formalistic understanding of the reduction of mathematics) but it has substantially influenced the character of mathematics. Namely, Boole–Schröder algebra has been founded as an abstract mathematical system, independent of its particular interpretations, by the method of deduction from a small number of premises. This method has become one of the main characteristics of mathematics, regardless of the quantitative or any other character of the matter being researched. However, the method itself is paradoxical if the system, which is to be founded, is the formal logic itself. Namely, mathematical deductions are unavoidably based on the laws of logic which are yet to be founded. Consequently, the result of purifying formal logic by its dearithmetization, was not a pure formal logic, but rather a theory of pure mathematics. In this sense, G. Boole was the father of pure mathematics and impure mathematical logic.[1] Here we end the survey of the period which originally considered formal logic as a special arithmetic, which in time came to be viewed as a special case of pure mathematics and which introduced the term mathematical analysis of logic, reminiscent of the term mathematical logic, if we read it as **mathematics of logic.**

This brings us to the new period in the development of formal logic and mathematics. The criticism of the infinitesimal calculus and the return of mathematics to the problem of its foundation in 1850s, clearly stated the need for the foundation of the arithmetic of real numbers. It is a period of critical movement in mathematics, a period when K. Weirstrass became a supreme mathematical authority ("Notre maître à tous", said Hermite), a period in which the intuition of space and time is being rejected as a basis of arithmetic. In this period, R. Dedekind arithmetized the continuum of real numbers in [4] and further posed the problem of the genesis of natural numbers, saying: "When I say that arithmetic (algebra, analysis) is only a part of logic, I already state that I consider the concept of number completely independent of the notions or intuitions of space and time, and, on the contrary, I consider it to

[1]Yet, some "purification" of formal logic did take place. The various interpretations of Boole–Schröder algebra showed that the logic of concepts and one-place propositional functions could be founded on the logic of propositions. Since this algebra eventually remained in the background of the logical researches and was not being frequently connected with the later logic of propositions and propositional functions, it was only recently that this important result was used to prove, in a simple way, the decidability of monadic logic in a student textbook (cf. [22] ch. 24.).

be a direct product of the pure laws of reasoning" cf. [5]. It is the beginning of the period which is not interested in a logic which looks for support in arithmetic. On the contrary, it looks for support of arithmetic and mathematics in logic.

However, this radical turning point was gradually prepared by adopting and developing one more aspect of the critical movement in mathematics. In the minds of the leading mathematicians at the end of nineteenth century, there is a clear-cut ideal of a mathematical theory which is derived from a small number of mathematical premises in accordance with the logical principles. This is the ideal which rejects intuition even as a means of demonstration. Some more radical thinkers, like G. Peano, realized that a deductive science, understood in this way, demands postulation and definition of its basic entities. He has remarkably pointed to the trickiness of the intuitive comprehension of the basic mathematical entities, by constructing a curve which fills a two-dimensional area [17]. But it also demands a radical cleansing of the deductive process from the influence of intuition. It can be formulated only within the framework of a symbolic language which is freed from the intuitive content of natural languages. Mathematics has already developed such a language to a great extent. However, Peano thinks that it is necessary (according to the new concept of the deductive science) to do something that has not been done even in mathematics: to formalize and symbolically describe the very arguments of mathematics. With this purpose in mind he designed a mathematical logic and used it in the formalization of mathematics [18, 19]. Due to this mathematical logic, a derivation of a conclusion from premises was replaced by a formal generation of the respective symbolic expression, from other expressions of that kind, by a quasi-algebraic process.

It seems as if this procedure was not essentially different from Boole's. Yet, it is. Boole's logic is an application of pure mathematics and the former (logic) is based on the latter (pure mathematics). Peano's logic is the basis of pure mathematics and here the latter is based on the former.

Hence, mathematics, freed from intuition, becomes a set of propositions of the form "p implies q", where p represents a conjunction of the postulated mathematical assertions (axioms) and q their quasi-algebraically (mechanically) derived consequence. (In a sense it was the beginning of Hilbert's formalism.)

The connection between this aspect of the critical movement with the one represented by Dedekind, who stated that the basic constituents of mathematical propositions (natural number, integer, rational number, real number) are logically definable,[2] draws attention to the possibility of p and q being purely logical propositions (propo-

[2]The construction of real numbers by Dedekind's cuts (i.e., the arithmetization of the continuum) should have shown the logical definability of real numbers by rational numbers. Logical definition of rational numbers by natural numbers is also to Dedekind's merit, as well as the reduction of the natural number concept to the logical concept of chain (by abstracting the nature of its elements). The possibility of the logically founded process of abstraction by the concept of the similarity of sets is due to Cantor. It is interesting that Dedekind's concept of chain (in [5]) anticipates Peano's axiomatization of natural numbers from [16]. However, each of these two works represents just one of the two above-stated aspects of the critical movement in mathematics. Thus, Peano postulates the principle of the mathematical induction by formally expressing it, while Dedekind has informally proved it as a consequence of purely logical characteristics of chains. Peano's axiomatization is

sitions made up of purely logical concepts). It means that mathematics, freed from intuition, could become a set of purely logical propositions of the form "p implies q". In other words, that mathematics could be just a part of logic. Joining these two aspects of the critical movement in mathematics and a consistent acceptance of such an attitude is to the merit of B. Russell.

In this way, a logicist period in the apprehension of the relations of formal logic and mathematics has been initiated, a period which considers arithmetic, and consequently all pure mathematics, as a part of logic and which accordingly introduces the term mathematical logic (in [18] for the first time, as far as the author knows) reading it as **logic of mathematics**.

At the end, it is necessary to mention that the joining of these two basic aspects of the critical movement (the joining which we identified with the beginning of logicism) had been carried out not only before Russell's [23] but even before [5, 16] (which represent just one of the two aspects). Logicism was already conceived in [6] by G. Frege, in 1879. In this work, mathematical argument was formalized in Peano's sense of the word, but with a thoroughness which had never been achieved by Peano. It was formalized in the form of the quantification theory (for the first time). It is here that the concept of the sequence of natural numbers was founded, by a method analogous to Dedekind's, in the context of the already formalized system of arguments (for the first time). It might seem strange then that we do not start the survey of this period with Frege. Nevertheless, it is in keeping with the mainstream of the development of these ideas. Frege's work met with complete misunderstanding and it remained unknown for 20 years (until B. Russell draw attention to its significance), and it became widely acknowledged only 50 years after it had been published. (It is interesting that in his honorary speech on the occasion of Frege's nomination for the associate professor in Jenna, E. Abbe predicted exactly such a sequence of events, saying that [6] will exercise considerable influence on mathematics but not immediately, because there will hardly be a person who could have an attitude toward such an original entanglement of ideas.)

The reasons for this misunderstanding are multifold. Frege's work falls into Boole's period. Frege did not know about Boole's work and could not explicitly point to the radical difference between his logicist and Boole's mathematical approach. Furthermore, the title of his work was understood (at the time it was published) in Boolean way, as an arithmetization of logic. From this point of view, Schröder gave a review of [6] in [25], by comparing Boole's and Frege's symbolic notation and he, naturally, concluded that Boole's one has advantages being taken over from arithmetic, but he wrongly claimed that the scope of Frege's logical analysis was not wider than Boole's. In Venn's review [28], the author came to the same conclusion, and so it is no wander that he wrote of Frege as "one of those instances of an ingenious man working out a scheme in this case very cumbrous one in entire ignorance that anything of the kind had ever been achieved before" [29, p. 415]. Frege could not but

not original. It can be found informally and in broader lines in Grassmann's [11] and, of course, in Dedekind [5], as quoted by Peano himself. He is one of the few mathematicians who noticed and publically promoted the importance of Grassmann's work, particularly in geometry.

regret that J. Venn, E. Schröder and others did not try to rewrite some formulae from the third chapter of his book in Boole's notation [7]. They would have immediately found out that it was not possible because Boole's logic, after being translated into the language of the theory of quantification, coincides with the logic of one-place predicates only, while Frege's system includes the complete first-order logic. In connection with this never made translation, it should be noted that P. Tannery in his review [27] of Begriffsschrift called Frege's replacement of the concepts of subject and predicate by that of function and argument fruitless and unfavorable. This is the replacement which founded the theory of quantification and which enabled the logic of one-place predicates to be simply extended to many places predicates (replacing in this way the clumsy theory of relations, which represents a respective extension in Boole–Schröder tradition). Realizing what was going on, Frege wrote an article in which he explained the main differences between his and Boole's system. However, the article, as well as its short version, was refuted by three journals. The fate of [6] was also the fate of [8] and of the first part of [9] (the second part of [9] was not published because the editor claimed the public was not interested in it). The first part of [9] was reviewed by G. Peano in [20], where he wanted to show the predominance of his own logic over Frege's. However, this marked the beginning of Frege's successful period. By his answer (in [10]) Frege causes Peano to acknowledge the significance of his logic and to thank him for the improvement of his own. Besides, due to Peano, B. Russell got to know Frege's work, and he began to advertise Frege's significance (realizing that the logicism he was aiming at had already been worked out by Frege). On the other hand, B. Russell delivered a severe blow to the whole Frege's program, by the famous "Russell's paradox" [24].

References

1. Boole G (1847) The mathematical analysis oflogic, being an essay towards a calculus of deductivereasoning. Macmillan, Barclay, and Macmillan, Cambridge
2. Boole G (1854) An investigation of the laws of thought on which are founded the mathematical theories of logic and probabilities. Macmillan, London
3. De Morgan A (1849) Trigonometry and double algebra. London
4. Dedekind R (1872) Stetigkeit und irrationale Zahlen. Vieweg, Braunschweig
5. Dedekind R (1887) Was sind und was sollen die Zahlen? Vieweg, Braunschweig
6. Frege G (1879) Begriffsschrift, eine der arithmetischen nachgebildete Formelsprache des reinen Denkens. Nebert, Halle
7. Frege G (1883) Über den zweck der begriffsschrift. In Sitzungsberichte der Jenaischen Gesellschaft für Medicin und Naturwissenschaft, JZN 16, pp 1–10
8. Frege G (1884) Die Grundlagen der Arithmetik. Eine logisch-mathematische Untersuchung uber den Begriff der Zahl. W. Köbner, Breslau
9. Frege G (1893) Grundgesetze der Arithmetik Begriffsschriftlich abgeleitet (Band I & II). Jena, H. Pohle, p 1903
10. Frege G (1896) Lettera del sign. g. frege alleditore. Rivista di Matematica 6:53–59
11. Grassmann H (1861) Lehrbuch der Arithmetik. Berlin
12. Hamilton WR (1837) Theory of conjugate functions, or algebraic couples; with a preliminary and elementary essay on algebra as the science of pure time. Trans R Ir Acad 17:293–422

13. Javons WS (1864) Pure logic or the logic of quality apart from quantity. London
14. Leibniz GW (1765) Oeuvres Philosophiques Latines Et Françoises de Feu Mr. de Leibnitz: Tire'es de Ses Manuscrits Qui Se Conservent Dans La Bibliothèque Royale a Hanovre. Eric Raspe, Amsterdam-Leipzig
15. Peacock G (1830) Treatise on algebra. Cambridge
16. Peano G (1889) Arithmetices principia. Turin
17. Peano G (1890) Sur une courbe qui remplit toute une aire plane. Math Ann 36:157–160
18. Peano G (1894) Notations de logique mathematique. Turin
19. Peano G (1895) Formulaire de mathematiques. Turin
20. Peano G (1895) Review of freges grundgesetze der arithmetik i. Rivista di matematica 5:122–8
21. Peirce CS (1880) On the algebra of logic. Am J Math 3:15–57
22. Quine WV (1950) Methods logic. Holt, New York
23. Russell B (1903) The principles of mathematics, vol 1. Cambridge University Press, Cambridge
24. Russell B (1967) A letter from B. Russell to G. Frege. In: van Heijnoort I (ed) From Frege to Gödel: a sourcebook in mathematical logic. Harvard University Press
25. Schröder E (1880) Review of freges begriffsschrift. Zeitschrift für Mathematik und Physik 25:81–94
26. Schröder E (1890) Vorlesungen über die Algebra der Logik (Exakte Logik). Leipzig
27. Tannery P (1879) Review of freges begriffsschrift. Revue philosophique 8:108–109
28. Venn J (1880) Review of freges begriffsschrift. Mind 5:297
29. Venn J (1881) Symbolic logic. Macmillan, London

Chapter 2
The McCulloch–Pitts Paper from the Perspective of Mathematical Logic

Tin Perkov

Abstract We analyze the McCulloch–Pitts seminal 1943 paper on a logical calculus for neural networks from the point of view of contemporary mathematical logic. Originally presented as a fragment of Carnap's Language II, the cited calculus is regarded in this chapter as a simple propositional temporal language with past tense modalities. We rewrite proofs of main results of the McCulloch–Pitts paper using this language.

Keywords Pitts McCulloch paper · Carnap's language II · Networks with cycles · Relative inhibition · Propositional temporal languages

2.1 Introduction

McCulloch and Pitts [3] use a logical language to mathematically model a notion of neural network. We omit neurobiological motivation and refer the reader to the cited paper for more details. This chapter treats neural networks as presented in [3] simply as abstract mathematical structures used as models of a logical language.

The authors use Carnap's Language II (cf. [1]), a higher order language of great expressive power, which covers basically anything one might need to say about relational structures, but they use actually a very small fragment of this language. In essence, it is a modal language of linear discrete sequential time and it will be presented as such in this chapter.

In Sect. 2.2 we rewrite the main result of [3], a formal description of conditions for excitation of a neuron in a network without cycles, in terms of modal logic. In Sect. 2.3 we discuss briefly one of several equivalence results of the cited paper, in this case, that the same formalization covers both absolute and relative notion of

This work has been supported by Croatian Science Foundation (HRZZ) under the project IP-01-2018-7459.

T. Perkov (✉)
Faculty of Teacher Education, University of Zagreb, Zagreb, Croatia
e-mail: tinperkov@gmail.com

S. Skansi (ed.), *Guide to Deep Learning Basics*,
https://doi.org/10.1007/978-3-030-37591-1_2

inhibitory synapses. In Sect. 2.4 we discuss networks with cycles, we extend the formalization to cover cycles of length 1 and conclude with a remark on difficulties of the general case.

2.2 Networks Without Cycles

A neural network may be regarded as a directed edge-labeled and vertex-labeled graph in which vertices are called *neurons*, each neuron is labeled by a natural number called its *threshold*, while each edge is labeled by its weight, which is either a natural number interpreted as the number of *excitatory synapses* between neurons which this edge connects, or $-\infty$, which is interpreted as the existence of an *inhibitory synapse* between these neurons. Neurons without predecessors are called *peripheral afferents*.

A temporal formalism is used in order to express conditions under which a fixed neuron of a network is excited. The language is given by

$$F ::= p_i \mid \blacklozenge F \mid F_1 \vee F_2 \mid F_1 \wedge F_2 \mid F_1 \wedge \neg F_2,$$

where p_i ranges over a set of propositional variables, each representing a neuron.

Well-formed formulas of the above language are called *temporal propositional expressions* (TPE). This is the terminology of [3], although used for corresponding formulas of Carnap's Language II. We will not go in detail on this language, but simply note that by an obvious translation we obtain a 1-1 correspondence between the present notion of TPE and the notion of TPE from [3].

Temporal propositional expressions form a fragment of the basic modal language in which the negation is not used freely, but only in conjunction with a TPE. We use $\blacklozenge$ instead of $\Diamond$, since the latter will be used in an extended language needed to formalize networks with cycles.

We use standard Kripke semantics, with a fixed frame, which is not a neural network as one might expect, but the set of natural numbers, intuitively interpreted as discrete linear time with a starting point 0, while the accessibility relation corresponding to $\blacklozenge$ is the immediate successor relation, i.e., $t \Vdash \blacklozenge F$ if and only if $t-1 \Vdash F$. In other words, $\blacklozenge$ is the previous-time modality, i.e., the inverse of well-known next-time modality, and thus often denoted in the literature as $\bigcirc^{-1}$, X^{-1} or $\blacklozenge^{-1}$, depending on a choice of notation for the next-time modality, but we simplify the notation since we only use past tense modalities.

The valuation of propositional variables is also not completely arbitrary. It is partly determined by a choice of neural network. The intended interpretation of p_i is that the corresponding neuron is excited at the current moment. While the valuation of propositional variables corresponding to peripheral afferents is arbitrary, a choice of it determines the valuation of other propositional variables. Namely, a neuron is excited if and only if the sum of weights of all edges from its predecessors which were excited at the previous moment is equal or greater than its threshold. The time unit in our discrete-time model is called the *synaptic delay*.

Definition 1 Let $\mathcal{N}$ be a neural network with n neurons, k of which are peripheral afferents. Let propositional variables $p_1, p_2, \ldots, p_k$ correspond to peripheral afferents and $p_{k+1}, p_{k+2}, \ldots, p_n$ to other neurons. A *solution* of $\mathcal{N}$ is a tuple $(F_{k+1}, F_{k+2}, \ldots, F_n)$ of temporal propositional expressions containing only $p_1, p_2, \ldots p_k$ such that $t \Vdash p_{k+i}$ if and only if $t \Vdash F_{k+i}$ for each i and at each moment t.

We say that a temporal propositional expression F is *realizable* if there is a neural network $\mathcal{N}$ with peripheral afferents corresponding to propositional variables occurring in F such that there is $m \in \mathbb{N}$ and there is a neuron represented by a propositional variable p such that $t \Vdash p$ if and only if $t \Vdash \blacklozenge^m F$ at each moment t.

A solution provides conditions for the excitation of all neurons in terms of peripheral afferents. Conversely, given a formula, a realization provides a network in which this formula (or its shift several time units back) is a condition for the excitation of a neuron.

Proposition 1 ([3], Theorems I and II) *Every neural network without cycles has a solution. Conversely, every TPE is realizable by a neural network without cycles.*

Proof Take any neuron which is not a peripheral afferent. It is excited at a moment t if and only if

$$t \Vdash \blacklozenge \left(\bigvee_{S} \bigwedge_{i \in S} p_i \wedge \bigwedge_{j \in I} \neg p_j \right), \tag{2.1}$$

where S ranges over subsets of the set of indices of predecessors such that the sum of weights of edges connecting predecessors from S with the neuron is equal or greater than its threshold, while I is the set of indices of predecessors connected by inhibitory synapses to the neuron.

The above formula is clearly a TPE (except if S is empty, in which case the neuron cannot be excited, so we can use some antitautology, e.g., $p \wedge \neg p$, which is a TPE, instead).

Any propositional variable in this formula which corresponds to a neuron which is not a peripheral afferent can be replaced with a TPE in the same way as above, thus obtaining again a TPE. Since the network is finite and without cycles, this procedure ends as desired.

The converse is proved by induction on the complexity of TPE. The base case is trivial. A TPE of the form $\blacklozenge F$, under the induction hypothesis that F is realizable, is realized by a network with an additional neuron with threshold 1 and one excitatory synapse from a realizing neuron of F to the additional neuron (see Fig. 2.1).

For a TPE of the form $F_1 \vee F_2$, we connect both a realizing neuron of F_1 and a realizing neuron of F_2, by one excitatory synapse each, to a new neuron with threshold 1. Thus the new neuron is excited at a moment t if and only if $t \Vdash \blacklozenge(F_1 \vee F_2)$ (see Fig. 2.2).

The case of a TPE of the form $F_1 \wedge F_2$ is similar, just using a new neuron with threshold 2 (Fig. 2.3).

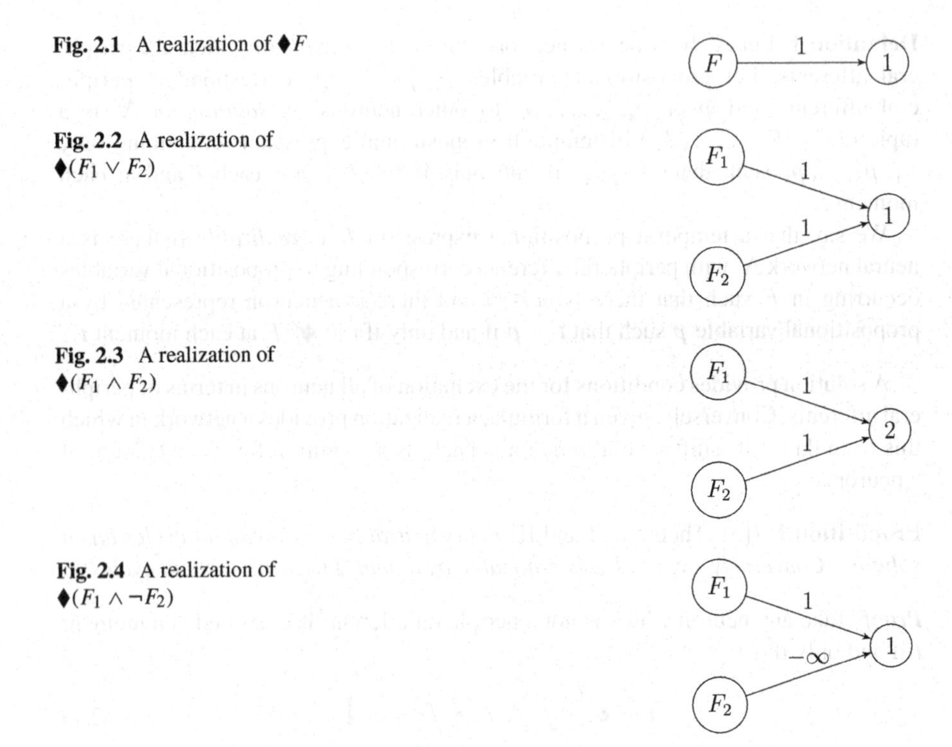

Fig. 2.1 A realization of $\blacklozenge F$

Fig. 2.2 A realization of $\blacklozenge(F_1 \vee F_2)$

Fig. 2.3 A realization of $\blacklozenge(F_1 \wedge F_2)$

Fig. 2.4 A realization of $\blacklozenge(F_1 \wedge \neg F_2)$

In the case a TPE is of the form $F_1 \wedge \neg F_2$, again we connect a realizing neuron of F_1 to a new neuron with threshold 1 by one excitatory synapse, while a realizing neuron of F_2 is connected to the new neuron by an inhibitory synapse, thus making its excitation expressed by $\blacklozenge(F_1 \wedge \neg F_2)$ (Fig. 2.4). □

2.3 Relative Inhibition

McCulloch and Pitts discuss the case absolute inhibition is replaced by relative inhibition, i.e., an inhibitory synapse does not necessarily prevent the excitation of a neuron, but raise its threshold. We will model this by labeling inhibitory edges of a neural network with negative integers instead of $-\infty$.

It turns out that relative and absolute inhibition are equivalent in the sense that Proposition 1 also holds for networks with relative inhibition and there is a natural transformation of a network with absolute inhibition to one with relative inhibition, and vice versa, realizing the same TPE, perhaps just in different time.

First, to show that the excitation of a neuron in a network with relative inhibition is also expressed by a TPE, we modify (2.1) to

$$t \Vdash \blacklozenge \left(\bigvee_S \left(\bigwedge_{i \in S} p_i \wedge \bigwedge_{j \in P \setminus S} \neg p_j \right) \right), \tag{2.2}$$

where P is the set of indices of predecessors and S ranges over subsets of P such that the sum of weights of edges connecting predecessors from S with the neuron is equal or greater than its threshold.

Note that Proposition 1 implies that TPE from (2.2) is realizable by a network with absolute inhibition. Conversely, a network with absolute inhibition is easily transformed into a network with relative inhibition, by replacing weights $-\infty$ with negative integers sufficiently large by absolute value to have the same effect in the network.

We omit a discussion of some other possible modifications in the assumptions on the notion of neural network, like extinction, facilitation, temporal, and spatial summation, which all can be similarly adapted to our modal framework.

2.4 Networks with Cycles

If we further generalize neural networks by allowing for directed cycles, the language of TPE is no longer sufficient, since we need to express looking an indefinite number of time units back in the past. To this purpose, we extend the language with an additional modality $\lozenge$, intuitively meaning "at some moment in the past". This is a well-known temporal modality often denoted in the literature by P or $\lozenge^{-1}$, if $\lozenge$ is chosen to denote the corresponding future modality. As before, we choose to avoid the inverse notation, so we extend our language to

$$F ::= p_i \mid \lozenge F \mid \blacklozenge F \mid F_1 \vee F_2 \mid F_1 \wedge F_2 \mid F_1 \wedge \neg F_2.$$

The truth clause for $\lozenge$ is: $t \Vdash \lozenge F$ if and only if $t' \Vdash F$ for some $t' < t$, which is the standard Kripke semantics with respect to the accessibility relation $>$.

McCulloch and Pitts first use cycles to model a so-called *alterable synapse*, i.e., an inactive synapse that becomes active if at some moment its initial neuron is excited, and accidentally (via a synapse from some other neuron) at the next moment its terminating neuron is excited, behaving after that like an ordinary excitatory synapse. By Theorem VII in [3], alterable synapses can be replaced by cycles. Namely, a minimal example which then can be used in the step of a full inductive proof we omit consists of two peripheral afferents and another neuron with threshold 1, whose excitation at a moment t is characterized by

$$t \Vdash \blacklozenge p_1 \vee \blacklozenge (p_2 \wedge \lozenge (p_1 \wedge p_2)),$$

i.e., either the first peripheral afferent, connected to the neuron by an excitatory synapse, was excited a moment ago, or the second one was, but with the condition

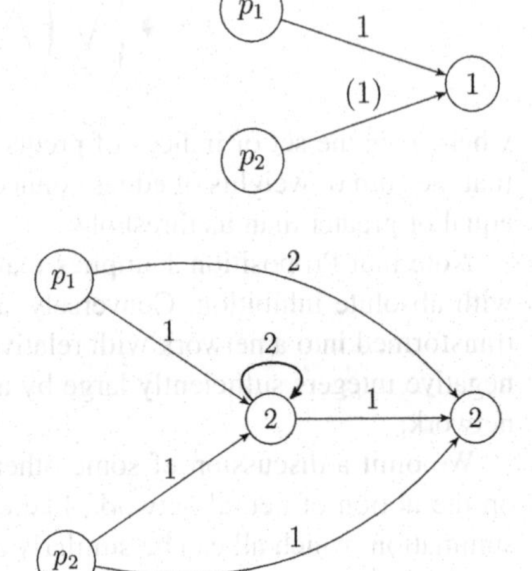

Fig. 2.5 A network with an alterable synapse (the weight in parentheses)

Fig. 2.6 A network with a cycle

that sometimes earlier both peripheral afferents were excited, thus activating the alterable synapse (see Fig. 2.5).

On the other hand, the same formula is realized by a network without alterable synapses, but with cycles, in this case, one additional neuron with threshold 2 and an edge of the weight 2 from and to itself, connected also by an excitatory synapse from both peripheral afferents and one toward the distinguished neuron, now with threshold 2, to which we also have edges from peripheral afferents, with weights 2 and 1, respectively (see Fig. 2.6).

Unfortunately, this discussion covers only cycles of length 1. The case of longer cycles is much more complex and exceeds the purpose of this article. McCulloch and Pitts discussion on this is brief and hard to follow, proofs are sketchy and no examples are given. Admitting these difficulties and also some suspicions in correctness of formulas in [3], Kleene [2] decides to independently study the problem and develops regular expressions as a natural framework for general neural networks.

References

1. Carnap R (1938) The logical syntax of language. Harcourt, Brace and Company, New York
2. Kleene S (1956) Representation of events in nerve nets and finite automata. In: Shannon C, McCarthy J (eds) Automata studies. Princeton University Press, pp 3–41
3. McCulloch WS, Pitts W (1943) A logical calculus of ideas immanent in nervous activity. Bull Math Biophys 5:115–133

Chapter 3
From the Linguistic Turn to the Cognitive Turn and Back Again

Marina Novina

Abstract The developments in the field of artificial intelligence are pointing out to complex nature of intelligence. The aim of this chapter is to show that to achieve a deeper understanding of intelligence, which is also the main task of Walter Pitts and Warren McCulloch, we need artificial intelligence, but we need psychology, neuroscience, and we also need philosophy. The complex nature of intelligence points to the need for turning back to evolution of understanding of intelligence that we find in definitions with which we operate and to history of the term itself. Moreover, it seems that the history of attempts to define intelligence indicate that its essence is scattered throughout many fields. However, the turn to the history of the meanings of the term itself can unravel the essence of intelligence. For this, we need philosophy. Perhaps we can say that we can grasp the essence of intelligence from mingling between the linguistic and the cognitive. For this, we need AI, psychology, neuroscience, and philosophy.

3.1 Introduction

The nature of intelligence is complex and therefore the meaning of the term "intelligence" is hard to grasp. This has implications for psychology, neuroscience, and Artificial Intelligence (AI). To achieve their own goals, these disciplines need not only to define but, moreover, to understand intelligence. There are many ways of defining this phenomenon. If any discipline in history has shown this, it is AI. Namely, as AI is developing we understand that intelligence is more than we thought or more than we supposed in our definitions. However, our definitions certainly are the starting points of our inquiries and points we return to. Furthermore, sometimes, as in case of intelligence, the so-called evolution of our understanding of a phenomenon is seen in variety of definitions. This variety is uncovering to us that the nature of the

M. Novina (✉)
Faculty of Philosophy and Religious Studies,
University of Zagreb, Zagreb, Croatia
e-mail: marina.novina@ffrz.hr

S. Skansi (ed.), *Guide to Deep Learning Basics*,
https://doi.org/10.1007/978-3-030-37591-1_3

phenomenon is complex and its essence is hard to grasp. However, if we turn to the term itself we see that from the metamorphosis of its meaning we can come closer to grasping the essence of intelligence and deepen our understanding of it. For this we need AI, psychology, neuroscience, and philosophy.

There were not many disciplines throughout history that challenged our understanding of intelligence as AI did. Hence, the developments in the AI field are pointing to complex nature of intelligence, and to the need of an adequate philosophy. In this sense, we can say that today is an interesting moment for psychology, neuroscience, and AI, which are called to turn back to philosophy for a deeper insight into the meaning of the term they operate with. In addition, this is an interesting moment for philosophy, which is called to turn to psychology, neuroscience, and AI to broader its understanding of intelligence. Even though there is no agreed-upon definition of intelligence, either in psychology or in philosophy, dispute about intelligence is still vivid. Moreover, the concept of intelligence is evolving and migrating from field to field. The history of this evolution, in the sense of attempts of defining the intelligence and in the sense of metamorphosis of the concept, is uncovering that the essence is scattered through many fields. To better understand intelligence and to come closer to the grasp of its essence, the main tasks Walter Pitts and Warren McCulloch wanted to achieve, we need to turn to philosophy. The aim of this chapter is to suggest how to grasp the essence of intelligence by turning to the history of the term "intelligence". Some parts of this history are and some are not well known outside of the context of philosophy. Hence, we will turn to these parts in order to come closer to our objective. This turn we will call "the turn for the missing pieces of the meaning". We will not open philosophical discussions on the nature of intelligence, or on the nature of definitions or relations between meanings and essence. Rather we will indicate why philosophy is essential for the developments in the field of AI, psychology, neuroscience, etc., and why it is essential for better understanding of intelligence. Furthermore, we will not offer a new theory of intelligence, or a new definition. However, we will show that throughout the history of attempts to define intelligence and through the history of metamorphosis of the term, we can indicate the essence of intelligence that has implications for the field of artificial intelligence.

3.2 Turning Back for Missing Pieces of the Meaning, or: Why Philosophy Matters?

Sometimes, to get the point, we need to reverse the story. We need to turn back for missing pieces of the meaning. Therefore, we start from the father of computing who marked the first step of the birth of artificial intelligence [19, p. 2]. Alan Turing started his famous paper *Computing Machinery and Intelligence* with the proposal to consider the question: Can machines think? He stated that to answer this question, he needed definitions of the meaning of the terms "machine" and "think" but concluded that "the definitions might be framed so as to reflect so far as possible the normal use of the words, but this attitude is dangerous" [23, p. 433]. Instead of the definitions of the terms, he proposed the "imitation game," and the today well-known concept of the "Turing test" was born. Thus, when John McCarthy coined the term "artificial intelligence" (1955) [13, p. 2], he did not just create the term for the new field of AI, he reopened the question and the problem Turing faced. In his book *Defending AI Research* McCarthy commented the so-called *Lighthill Report*.[1] One of the basic ideas of the *Lighthill Report* is that "the general structure of the intellectual world is far from understood, and it is often quite difficult to decide how to represent effectively the information available about a quite limited domain of action even when we are quite willing to treat a particular problem in an ad hoc way" [16, p. 28]. Lighthill concluded that AI has not contributed to applications in psychology and physiology. McCarthy stated that Lighthill ignored the possibility that AI has goals of its own and showed that AI has contributed to other subjects. For McCarthy, *Lighthill Report* is giving up on science. For him, AI is promising, but "a very difficult scientific study, and understanding intelligence well enough to reach human performance in all domains may take a long time between 5 years and 500 years. There are fundamental conceptual problems yet to be identified and solved, so we can't say how long it will take" [16, pp. 29, 36]. From Turing's words quoted at the beginning of this section, we can conclude that he would agree with McCarty regarding the fundamental conceptual problems.

Furthermore, commenting on John Haugeland's book *Artificial Intelligence: The Very Idea*, McCarthy said that Haugeland got many things wrong but that he was right about two things. The first is the polarization between the scoffers and the boosters of AI regarding the self-assurance of both sides about the main philosophical issue. Second, he is right about the abstractness of the AI approach to intelligence. For, as McCarthy says, "we consider it inessential whether the intelligence is implemented by electronics or by neurochemical mechanisms or even by a person manipulating pieces of paper according to rules he can follow but whose purpose he doesn't understand" [16, p. 40]. Moreover, for McCarthy, the book contents show that many issues raised by philosophers from Aristotle, Hobbes, and Descartes to Leibniz and Hume are alive today, but in an entirely different technological context. Even so, "it's hard to trace any influence of this philosophy on present AI thought or even to argue that reading Hobbes would be helpful. What people are trying to do today is almost entirely determined by their experience with modern computing facilities rather than by old arguments however inspired" [16, p. 41]. We do not know what Turing would say on this comment, although his famous paper was published in *Mind*. However, we think that Turing was right that the main problem is the definition, but, unlike McCarty, we think that to solve this problem we need philosophy. To understand intelligence well enough to try to reach human performance demands interdisciplinary and complementary approach to a phenomenon.

Now, just as thinking was the epitome of intelligence and logic considered as something connected with the laws of thinking, and just as it was, therefore, natural

[1] Name used for the paper "Artificial Intelligence: A General Survey" by James Lighthill, published in Artificial Intelligence: a paper symposium in 1973. It is the proposal to end support for AI research

that AI started out with logic [19, p. 12], it is even more natural than AI, as any other science, needs philosophy for solving fundamental conceptual problems. Perhaps it is "inessential whether the intelligence is implemented by electronics or by neuro-chemical mechanisms or even by a person manipulating pieces of paper according to rules he can follow but whose purpose he doesn't understand" [16, p. 40], but it is essential to know what intelligence is. The first fundamental problem concerns the concept "intelligence" and we should consider its nature more closely. How do we define intelligence and can the meanings of the term "intelligence" indicate an essence of intelligence?

Without doubt, in routine scientific practice, it is not easy to trace philosophical influence on the discipline with which we are engaged. However, we should be aware that philosophy is mother of the sciences on the one side and on the other side that for routine scientific practice we need concepts. Concepts are main connections between human activities, culture, scientific disciplines and philosophy, and it is the task of philosophy to discuss and understand the nature of these connections. Moreover, concepts are necessary tools for every science, starting points of every scientific inquiry. We can generalize from the history of philosophy and say that "a discipline remains philosophical as long as its concepts are unclarified and its methods are controversial. Some may say that no scientific concepts are ever fully clarified, and that no scientific methods are ever totally uncontroversial. If that is true, all that follows is that there is always a philosophical element left in every science" [14, p. 4].[2]

However, concepts we use have a history; we can say that their meanings evolve and that they migrate between disciplines. There are many patterns according to which concepts we use evolve

> [...] but usually they become more and more general. However, this process is not a linear one. It can have many side-branches and diverse ramifications. The objective of science is to explain each specific phenomenon, but to explain a phenomenon in science means to put it in a more general conceptual scheme. Moreover, in this generalization process old concepts are not eliminated but engulfed by new ones as their "special cases". We could risk the statement that the generalization of conceptual schemes determines an "arrow of time" of scientific progress. During every such metamorphosis they changed their meanings by adapting themselves to a new environment. The history of ideas is full of instances of these processes. But it would be naive to think that they belong only to the past [12, pp. 247–248].

Psychologist Robert J. Sternberg said that as anyone who has seriously studied the history of any country knows, there is not one history of a country but many histories. Similarly, he said, there is no one history of the field of intelligence, but many histories that depend on who is doing the telling [21, pp. 3–4]. The same applies to the concept of intelligence. The nature of intelligence is still a big question, and Sternberg, in his book *Metaphors of Mind. Conceptions of the Nature of Intelligence* [20], said that metaphors, and there are many metaphors of intelligence, serve as the foundations for theories of intelligence. Again, we do not have an answer to the

[2]For the studies about the conceptual parallels between philosophy and cognitive science see [17]. For the philosophical introduction to AI see [4, pp. 4–10], and for the philosophy of AI see [3].

question what intelligence is nor agreed definition of intelligence. It depends on who is doing the telling.

Therefore, we need to turn back to history on two levels. First, we need to turn back to history of the metamorphosis of understanding of nature of intelligence that we see in definitions we use. Second, we need to turn back to history of the term itself. To neglect the history is to give up on a part of the meaning and to reduce our understanding of a phenomenon. We might say that science helps us to understand a phenomenon in some new way, but science starts with the definition of the concept it uses. From the concept, using definitions, we turn to inquiry and from inquiry we turn back to the concept. However, to know the history of the concept and the meanings of a term itself is to know the phenomenon deeper. This means to come closer to the essence that is scattered throughout many fields. For this, we need philosophy. We cannot open the philosophical discussions on these questions but we can indicate that reading of some philosophical texts on this subject would be very useful, but to have philosophy of an adequate depth it would be precious. In this sense, since AI uses definition of intelligence to turn to its own inquiry, it would be very useful for the goals of AI to have an appropriately deep philosophy. The inquiries of AI researchers, particularly in the field of deep learning, help us to understand intelligence better, which is good for psychology, neuroscience and philosophy. Sometimes in this turn from the linguistic to the cognitive and back we find forgotten or hidden pieces of meaning. Perhaps we can say that we can grasp the essence from mingling between the linguistic and the cognitive. Certainly, we are achieving better understanding of intelligence. This would surely be something that Walter Pitts and Warren McCulloch, with whose seminal paper, *A Logical Calculus of Ideas Immanent in Nervous Activity*, the neural network history begins, would have wanted [19, pp. 3–5]. Our first turn is the turn to the question of what intelligence is or to the attempts at defining intelligence.

3.3 The Definition, or What Is Intelligence?

Throughout human history intelligence was understood in various ways, as the capacity for abstract thought, understanding, communication, planning, learning, reasoning, and problem-solving. Today "intelligence" is the term we apply in human but also in non-human context and, consequently, it is studied in the context of human species and in animals and plants [11, p. 7]. The term evolves definitions multiply and epistemologies change. The term "intelligence" is widely used but the question what intelligence is still puzzles the minds of many philosophers, scientists, and psychologists.

Since intelligence is inquired first and foremost in human context, we naturally turn to psychology for the answer to the question what intelligence is. However, we find that in psychology there is no agreed-upon definition of intelligence. Moreover, as Strenberg notes: "although all fields of psychology are perceived through ideological lenses, few fields seem to have lenses with so many colors and, some might argue, with so many different distorting imperfections as do the lenses through which are seen the field of intelligence" [21, p. 4]. The reason for this lies in the fact that intelligence is the construct on which psychologists base their models, but at the same time, the fundamental question is how to conceive intelligence.

In psychology, there are two well-known attempts to define intelligence. The first is based on the famous study of experts' definitions of intelligence. It was done by the editors of the *Journal of Educational Psychology* in 1921 (*Intelligence and its measurement*) and known as *The 1921 Symposium*. Experts were asked to address two issues. First, what they conceived intelligence to be and how it could be best measured by group tests? Second, what would be the next most crucial research step? Fourteen experts gave their views on the nature of intelligence. They offered 11 definitions. Thorndike said that intelligence is the power of good responses from the point of view of truth or facts. Terman said that it is the ability to carry on abstract thinking. Freeman said that it is sensory capacity, capacity for perceptual recognition, quickness, range or flexibility of association, facility and imagination, span of attention, quickness or alertness in response. Colvin saw it as something learned or ability to learn to adjust oneself to the environment, and Pintner as the ability to adapt oneself adequately to relatively new situations in life. Henmon defined intelligence as the capacity for knowledge and knowledge possessed, while Peterson saw it as a biological mechanism by which the effects of a complexity of stimuli are brought together and given a somewhat unified effect in behavior. Thurstone defined intelligence as the capacity to inhibit an instinctive adjustment, the capacity to redefine the inhibited instinctive adjustment in the light of imaginably experienced trial and error, and the capacity to realize the modified instinctive adjustment in overt behavior to the advantage of the individual as a social animal. While Woodrow thought it is the capacity to acquire capacity. For Dearborn intelligence was the capacity to learn or to profit by experience and for Haggerty it was sensation, perception, association, memory, imagination, discrimination, judgment, and reasoning. Other contributors to the symposium did not provide clear definitions of intelligence [21, p. 6].

Another well-known attempt to define intelligence was done in 1986 by two leading figures in psychological research on intelligence. Douglas K. Detterman and Robert J. Sternberg tried to update the 1921 symposium. They

> Solicited two dozen brief essays by experts in the field of intelligence, who were asked to respond to the very same questions that were posed to the experts in the 1921 symposium. The idea was to address the issues raised in a way that might reflect any progress that had been made from the beginning to the ending of this century [15, p. 19].

The two dozen answers were published the same year in book titled *What Is Intelligence? Contemporary Viewpoint on Its Nature and Definition*. Sternberg and Berg made the comparison of the 1921 and 1986 attempts to define intelligence and came to three insights. First, attributes such as adaptation to the environment, basic mental processes, higher order thinking (e.g., reasoning, problem-solving, and decision-making) were noticeable in both symposia. Therefore, they concluded that there was at least some general agreement across the two symposia regarding the nature of intelligence. Second, central themes occurred in both symposia, but main question was: Is intelligence one thing or is it multiple things, and how broadly should intelligence be defined? Third, despite the similarities, difference was metacognition, conceived of as both knowledge about and control of cognition. It played a prominent role in the 1986 symposium, but almost no role at all in 1921. The 1986 symposium also placed a greater emphasis on the role of knowledge and the interaction of mental processes with this knowledge [21, p. 7]. Their general conclusion was:

> The field of intelligence has evolved from one that concentrated primarily upon psychometric issues in 1921 to one that concentrates primarily upon information processing, cultural context, and their interrelationships in 1986. Prediction of behavior now seems to be somewhat less important than the understanding of that behavior, which needs to precede prediction. On the one hand, few if any issues about the nature of intelligence have been truly resolved. On the other hand, investigators of intelligence seem to have come a rather long way toward understanding the cognitive and cultural bases for the test scores since 1921 [22, p. 162].

Based on the gathered definitions of intelligence on two symposia Sternberg offered an integrative framework for understanding the conceptions of intelligence. The proposed framework branches understanding of intelligence in three contexts. First, *individual* context that includes: (A) Biological level: 1. across organisms (between species (evolution), within species (genetics) and between-within interaction); 2. within organism (structure, process, and structure-process interaction); 3. across-within interaction. (B) Molar level: 1. cognitive (metacognition (processes, knowledge, and process-knowledge interaction), cognition (processes (selective attention, learning, reasoning, problem-solving, decision-making), knowledge and process-knowledge interaction) and metacognition-cognition interaction); 2. Motivational (level (magnitude) of energy, direction (disposition) of energy, and level-direction interaction). (C) Behavioral level: 1. academic (domain-general, domain-specific, and general-specific interaction); 2. social (within-person, between-persons, and within-between interaction); 3. Practical (Occupational, Everyday Living, Occupational Everyday Living Interaction); (D) Biological-molar-behavioral interaction. Second, *environmental* context that includes: (A) Level of culture/society and (B) Level of niche within culture/society which booth branch on: 1. demands; 2. values and 3. demands-values interaction. (C) Level × sublevel interaction. Third, *individual-environment interaction* [22, pp. 4–5].

Charles Spearman (1904) proposed well-known g or general factor of intelligence (the stock of energy in the nervous system) and s or specific factor of intelligence (the structure of some particular area of group of neurons). "The general mental acts and is constant for the individual. It consists of the energy which is at the disposal of the whole brain. The specific factor is the structure of some particular area or group of neurons in the brain" [8, p. 260]. However, this did not solve the problem of the definition of intelligence; it only confirmed complexity of its nature [8, p. 260]. Therefore, Sternberg's framework is a good outline of the complex nature of intelligence. This complexity certainly indicates why we lack agreed-on definition of intelligence. In addition, with developments in AI, this complexity is a good confirmation of need for philosophical insights in ways of defining a phenomenon and for understanding of intelligence. Without doubt, for these requests it would be useful to read philosophers as Plato, Aristotle, Locke, Frege, Rusell, etc. that were discussing nature of definitions in their texts. Furthermore, the complex nature of intelligence indicates why Sternberg concluded that the basis for its understanding is metaphor. In that sense, it would be useful to study philosophical discussions on metaphors, analogy or links between words and objects, and in that sense Plato, Aquinas, Frege, Kripke, and Quine can be useful.

Some psychologists recognize this philosophical background. Jagannath P. Das starts his inquiry of the history of the definition of intelligence with the indication that we can trace attempts to define intelligence back to Aristotle, who divided mental functions into the cognitive (cybernetic) that are essentially thought processes and the orectic (dynamic) that comprise emotional and moral aspects.[3]Further, he says that similar dichotomy can be found in the Hindu system of logic where "Purusa, literally, the male energy, is thought which acts as a catalyst for Prakriti, the female energy, which is emotion and action. The former is passive, the latter is active" [6, p. 1]. Then he notices that we can recognize the same distinction today and concludes that "currently, we separate intelligence from personality or the cognitive from the affective domain, although such separation is recognized to be impossible at a practical level" [6, p. 1]. Das points out well that contemporary attempts to define intelligence have philosophical background and this background needs to be studied if we want to gain better understanding of intelligence.

Has intelligence evolved and how much, as we read in the 2015 Goldstein, Princiotta and Naglieri *Handbook of Intelligence: Evolutionary Theory, Historical Perspective, and Current Concepts*, depends on its definition, but our understanding of intelligence certainly has evolved and will grow as different disciplines put their forces together. We still do not have complete understanding of intelligence nor agreed-on definition of it, but the term is alive. Maybe as Lanz said the term "intelligence" stirs up more trouble than it can help to soothe [15, p. 21], but we think that from the meaning of the term we can derive the essence of intelligence. We can say that we are turning to the history of the term "intelligence". Some parts of this history are not well known outside of the context of philosophy.

3.4 The Term "Intelligence"

Expert in the field of intelligence themselves are aware that behind the definitions of intelligence and behind the concept with which they operate there is a philosophical thought and a philosophical term. Both is evident from their texts devoted to the nature and the concept of intelligence. Therefore, for example, Goldstein turns back to the meaning of the term "intelligence". He notes that the roots of the term "intelligence" are in the Latin verb *intelligere* and that in turn has roots in *interlegere*, and that the "form of this verb, *intellectus*, was the medieval technical term for someone with a good understanding" [11, p. 3]. This was the translation of the Greek term "nous".

[3]It is true that we can find this kind of distinction, although the term "cybernetic" is not quite appropriate. The distinction between the cognitive and the affective can be attributed to Plato.

Nous, however, was strongly linked to the metaphysical, cosmological theories of teleological scholasticism, including theories of the immortality of the soul and the concept of the active intellect. Its entire approach to the study of nature, however, was rejected by modern philosophers, including Francis Bacon, Thomas Hobbes, John Locke, and David Hume, all of whom preferred the word "understanding" in their English philosophical works. [...] The term intelligence, therefore, has become less common in the English language philosophy but has been widely adopted in contemporary psychology, absent the scholastic theories which it once implied [11, p. 3].

Goldstein well notices that contemporary uses of the term have lost some of the pieces of the meaning of the term itself. But these missing pieces can be important for our understanding of intelligence and have implications for the fields that operate with the concept of intelligence. Therefore, we turn back for those missing pieces of the meaning. Once more, we will reverse the story.

The term "intelligence" has its roots, as Goldstein noticed, in Latin, which through the forms of the term "intellect," that is translation of Greek term "nous," connects English, Latin, and Greek language. However, as Goldstein indicated something is missing from English usage of the term "intelligence," but it is not only some scholastic theory. Namely, in Latin language the term "intellectus" is noun use of past participle of "intelligere" that was translation of the Greek term "nous". The Latin term was probably translated from Aristotle, but in philosophy it was introduced by Anxagoras from its pre-philosophical usage. It should be noted that English language does not have a convenient verb for translation and it cannot cover the various activities of the intellect, as the Latin *intelligere*. As Kenny writes

To correspond to the Latin verb one is sometimes obliged to resort to circumlocutions, rendering *actu intelligere*, for example, as "exercise intellectual activity". An alternative would be to use the English word "understanding", in what is now a rather old-fashioned sense, to correspond to the name of the faculty, *intellectus*, and to use the verb "understand" to correspond to the verb *intelligere*. In favor of this is the fact that the English word "understand" can be used very widely to report, at one extreme, profound grasp of scientific theory ("only seven people have ever really understood special relativity") and, at the other, possession of fragments of gossip ("I understand there is to be a Cabinet reshuffle before autumn"). But "understand" is, on balance, an unsatisfactory translation for *intelligere* because it always suggests something dispositional rather than episodic, an ability rather than the exercise of the ability; whereas *intelligere* covers both latent understanding and current conscious thought [14, p. 41].

What intelligere covers is beast seen at Thomas Aquinas.[4] As Jason T. Eberl[5] outlines, in Aquinas we notice that the intellect should be understood as an essential capacity (not singular capacity, and not more than one intellect for each person, but one intellect that engages in various types of operations) of a human soul. Since

[4]One of the most influential scholastics and philosopher who is direct connection to early Christian philosophers and Greek philosophical thought, just as Aristotle is connection to thought of Presocratics. He is important to us not only because in his texts he gives an overview of the history of arguments of thinkers before and in his age on questions he dealt with, but because he was systematic, logically clear, and the Latin he used was precise the Latin used on first Universities.

[5]In this passage, we will use Eberl overview of Aquinas questions 79, 84–89 from *Summa theologiae* but for the start of deeper study we recommend questions [1, Sth. I. q. 75–89].

Aquinas is accepting Aristotle's hylomorphic understanding of the soul, he does not identify the soul as Plato with the intellect, nor does he use the term, like post-Cartesians, to refer to the "mind" or "consciousness". In that sense, the non-rational animals are conscious and capable of certain degree of cognitive activity, they have minds, but they do not have *intellectus*. But Aristotle and Thomas make the further distinction between *intellectus possibilis* and *intellectus agens* that are two functions of intellect. The possible intellect is like *tabula rasa*, but it has a cognitive architecture that makes it possible to comprehend the intelligible forms it receives. It is analogue to Chomsky's thesis that human beings are born with an innate language-learning ability (capacity to learn). However, the intellect abstracts the intelligible form from its sense-perception of individual human beings and this abstraction is accomplished by active (agent) intellect (a creative intellectual light; an ability that belongs to the individual thinker). There is no intellectual cognition without objects perceived through senses (phantasm). Furthermore, the intellect uses the received intelligible forms and acquires additional knowledge through reasoning (something the intellect does). In confronting a particular object, intellect has three primary operations. First, the intellects comprehension of intelligible form is done by abstraction (conceptual striping away), and in this sense intellection differs from sensation as *seeing as* from *seeing*. For instance, intellection consists in apprehending this person *as human*, as opposed to perceiving *this person*. However, all human seeing intrinsically involves *seeing as*. Second, the intellect can "compound and divide" comprehended concepts, and gain deeper understanding (comprehend essential properties and formulate propositions) or create fictional objects. Finally, the intellect can reason, that is, engage in syllogistic inferences. The reasoning takes various forms, which depends on object and purpose. In this sense, Aquinas distinguishes higher from lower reason, that is, reasoning about eternal objects with the goal to attain wisdom, and reasoning about temporal objects. These forms of reasoning form speculative reasoning that aims at discovering what is the case, while practical reasoning aims at determining what ought to be done (moral reasoning). Now, since the human intellect has the innate structure, both the speculative and the practical reasoning include first principles. The first principles of practical reasoning are background everyday premises, like the principle of non-contradiction. The most fundamental principle in practical reasoning is "good is to be done and pursued, and evil is to be avoided" and the main feature of cognitive architecture required for moral reasoning is *synderesis*.[6] that is fundamental disposition of the intellect to reason practically. This conscious application of first principles for Aquinas is conscience. Finally, the intellective soul knows itself, but by means of its activity of intellectual abstraction, that is the self-consciousness or self-knowledge is by-product of intellectual cognition [7, pp. 100–109]. The self-awareness is "an intimate internal experience of myself as an existing individual, concretely present to myself in my acts" [5, p. 73].

Therefore, *intelligere* covers all characteristics of *intellectus* that is translated by the English term "intellect". The intellect "is the capacity for understanding and

[6]For detailed overview of human acting on moral principles see whole book and for the indicated term [24, pp. 110–101, 205].

thought, for the kind of thinking which differentiates humans from animals; the kind of thinking which finds expression especially in language, that is in the meaningful use of words and the assignment of truth-values to sentences" [14, p. 41]. However, as we have seen, for Aquinas the hylomorphic character of human nature is reflected in the mode of human cognition. The gap between immaterial intellect and material realities, Aquinas bridges the gap between immaterial intellect and material realities with a complex psychological process of dematerialization that is accomplished by a hierarchy of cognitive powers, each grasping a different aspect of experienced reality [5, p. 10]. Human intellect is naturally directed toward the "essences that are in material objects, and it depends on the senses for access to such objects. Thus the senses are not the obstacle, but the vehicle, for human intellectual cognition" [5, p. 9]. The operation of intellectual thinking occurs when the possible intellect is "informed" by the intelligible form illuminated by the active intellect. In this moment, my possible intellect is actually formally united with the essence in that individual object. This is the moment of understanding. Our intellectual attention is turned toward the objects of perceptions (phantasms), so that sense and intellect are unified in my *seeing* as *seeing as*. "This 'turn' secures a unified experience of the world in which sense and intellect cooperate, in consonance with the hylomorphic soulbody unity that is the human individual" [5, p. 22].

However, those characteristics of *intelligere* are not only products of mediaeval philosophical theory of mind and Aquinas discussion with philosophical ancestors. Aquinas "intelligere" is the precise translation of the Greek term "nous". *Intelligere* was carrying the essence of the meaning that was present in the first usage of the term "nous" in pre-philosophical period. The analysis of the meaning and etymology of the term "nous" used in pre-philosophical period was made by K. von Fritz and it supports this kind of reasoning. Since there is no accepted etymology of the term "nous," von Fritz started his analysis with the discussion of Homer. He pointed out that Joachim Boehme distinguished three main meanings of the term "nous": "(1) 'Seele als Träger seelischer Erlebnisse', which may perhaps be translated as 'the soul as an organ of experiences'; (2) 'Verstand', which can be either 'mind' or 'intellect' or 'intelligence' or 'understanding'; (3) 'Plan', which is 'plan' or 'planning'" [9, pp. 80–81]. Furthermore, von Fritz stated that in order to clarify these distinctions Bruno Snell "pointed out that in the first case *nous* means an organ, in the second and third case the function of this organ but with the further difference that in the second case it means the function as such while in the third case it means the function at a given moment. The second meaning, then, would correspond to the present tense of a verb, the third to an aorist" [9, p. 81]. The detailed analysis led von Fritz to conclude that Schwyzer's proposal of etymology is far more correct and the term is probably derived from a root meaning "to sniff" or "to smell".

It is quite true, writes von Fritz,

> that in Homer *noein* appears more often connected with the sense of vision than with any other sense. But this need not always have been the case. The most fundamental and most original meaning of *noein* in Homer seemed to have been "to realize or to understand a situation." The most primitive case in which a situation becomes important is that in which there is

danger or where there is an enemy nearby. The most primitive function of the *nous*therefore would have been to sense danger and to distinguish between friend and enemy [9, p. 93].

Further, von Fritz presupposes

that in a very early stage of human development the sense of smell played a preponderant part in this function. One has only to point to the fact that even in our present day language we still speak of "smelling a danger." With the development of a higher civilization the sense of smell would naturally have been more and more replaced by the sense of vision. But the understanding of a situation remained nevertheless distinguished from the vision or even recognition of an indifferent object. It is not impossible that the emergence of this new concept of a purely mental function was greatly favored by the fact that the original connection of this function with the sense of smell receded more and more into the background, as the sense of vision became more preponderant in importance, and finally was completely forgotten [9, p. 93].

It is interesting to look back on Aquinas understanding of intellect and notice that for Aquinas "verbs like *percipere* and *experiri* serve as general verbs of cognition (like *intelligere* or *cognoscere*) to denote sensory or intellectual operations or some cognitive operation in general, but they carry an additional connotation of the objects intimate presence to the intellect" [5, p. 73].

At the beginning of further inquiry of the uses of the term by Presocratic philosophers von Fritz summarizes derivative meanings of the terms "nous" and "noein" that can be found already at Homer. Therefore, the term "nous": (1) sometimes implies the notion of a specific reaction of a person to a specific situation and can be thought as a specific attitude; (2) can be thought as planning (to escape from dangerous situation or to deal with the situation); a volitional element enters into the concept; (3) can be thought as something that remains in the purely intellectual field; the realization of the meaning of a situation or the deeper insight into its real nature; (4) penetrates beyond the surface appearance and discovers the real truth about the matter ("intuitive element"); (5) "makes far-off things present" (designate the imagination by which we can visualize situations and objects which are physically or temporally distant); and (6) indicates a certain amount of reasoning (can engage in syllogistic inferences) [10, pp. 223–225]. Therefore, from this, it seems that mentioned derivative meanings of the Greek terms are present in Aquinas understanding of *intelligere* from which term "intelligence" arrives. Moreover, it seems that those terms uncover the essence of intelligence. This essence, however, has clear implications for contemporary inquiries of intelligence, especially for the AI field.

3.5 The Essence of Intelligence and AI

From this short insight in the history of evolution of our understanding of intelligence and attempts of its defining, we can conclude that we can distinguish various types of definitions and that they, as philosophical discussions indicate, are of different types and have different purposes. However, this also indicates that the essence of intelligence is scattered throughout many fields or disciplines. This has direct implications for all the sciences. This is what Turing noticed and therefore proposed the "imitation game" that would correlate to IQ test that postulate parameters of intelligence without the answer on the question what intelligence is, or which of the definitions, in philosophical sense, would be real definition of intelligence. In that sense, we could say that intelligence remains a mystery. Nevertheless, indicated framework of understanding and studying of intelligence indicate that there is something that all definitions have in common. That is intelligence itself, which can be understood as g factor, Holy Grail as Margaret Boden called it, that AI cannot grasp. It seems that in spite of great developments in the field of AI that human level of AI is not within sight [2, p. 56]. As Sternberg's framework and developments in the field of AI indicate, it seems that intelligence includes more than logical reasoning. Language, creativity and emotions are challenges for the AI field, but intentionality, making moral decisions, intuition, and dreaming are also waiting around the corner to be challenged.

Writing on intentionality, John R. Sarle says that "anyone who tries seriously to follow out the treads the Network will eventually reach a bedrock of mental capacities that do not themselves consist in Intentional states (representations), but nonetheless form the preconditions for the functioning of Intentional states" [18, p. 143]. This background is hard to demonstrate, but he thinks that is pre-intentional: a set of nonrepresentional mental capacities that enable all representing to take place [18, p. 143]. It seems from this that the intelligence can be seen as some kind of background behind all of this challenges. But, Boden asked: what if AI equals human performance?

> Would they have real intelligence, real understanding, real creativity? Would they have selves, moral standing, free choice? Would they be conscious? And without consciousness, could they have any of those other properties? These arent scientific questions, but philosophical ones. [...] We need careful arguments, not just unexamined intuitions. But such arguments show that there are no unchallengeable answers to these questions. Thats because the concepts involved are themselves highly controversial. Only if they were all satisfactorily understood could we be confident that the hypothetical AGI would, or wouldnt, really be intelligent. In short: no one knows for sure. Some might say it doesnt matter: what the AGIs will actually do is whats important. However, our answers could affect how we relate to them [...] [2, p. 119].

We tried to show that for the clarification of the question of intelligence we need AI, psychology, neuroscience, but also philosophy. However, just like Boden, we do not have unequivocal answers, but are suggesting that some answers are more reasonable than others [2, p. 120].

Therefore and furthermore, we think that the essence of this background can be grasped from mingling between the linguistic and the cognitive and that it is recognizable in the meanings of the term. This essence is sometimes over the history of metamorphosis and migration of the concept overlooked, forgotten, but implications of this essence remain present in today's challenges of AI. Therefore, to grasp the essence of intelligence we have turned back for the missing pieces of the meaning, to the history of the term itself. The look in this history has shown that English language cannot cover the various abilities of intellect expressed by the Latin *intellectus* and *intelligere*. However, a deeper insight in the mediaeval understanding of those terms in Aquinas philosophy of mind has shown that this understanding is not the product of certain philosophy but an attempt to uncover the essence of this mysterious background that we call intelligence. Moreover, the analysis of the Greek term "nous" and its etymological root shows that this essence was present in pre-philosophical uses of the term. We have indicated several characteristics of the term "intelligence" in Latin and Greek usage. From these characteristics, the meanings, it seems that the essence of intelligence includes dispositions for knowing the truth, for abstract and logical thinking and for language learning, but also includes senses, emotions, moral behavior, creativity, self-knowledge, intuition, and intentionality. Moreover, it seems that the essence of intelligence is not reachable out of specifically human cognitive structure. The implication for AI is clear: intelligence cannot be artificial. However, AI field of inquiry, especially deep learning, is very useful and worthy field of study because it, as Boden noticed, can illuminate the nature of real minds [2, p. 120]. That is something that Walter Pitts and Warren McCulloch would be satisfied with.

References

1. Aquinas T (1988) Summa theologiae. Editiones Paulinae, Milano
2. Boden MA (2016) AI its nature and future. Cambridge University Press, Cambridge
3. Carter M (2007) Minds and computers: an introduction to the philosophy of artificial intelligence. Edinburgh University Press, Edinburgh
4. Copeland JB (1993) Artificial intelligence: a philosophical introduction. Wiley-Blackwell, Oxford
5. Cory TS (2014) Aquinas on human self-knowledge. Cambridge University Press, New York
6. Das JP, Kirby JR, Jarman RF (1979) Simultaneous and successive cognitive processes. Academic Press, New York
7. Eberl JT (2016) The Routledge Guidebook to Aquinas Summa Theologiae. Routledge, London and New York
8. Freeman FN (1925) What is intelligence? School Rev 33(4):253–263
9. Fritz K (1943) ΝΟΟΣ and noein in the homeric poems. Class Philol 38(2):79–93
10. Fritz K (1945) ΝΟΟΣ, noein, and their derivatives in pre-socratic philosophy (excluding anaxagoras): Part i. from the beginnings to parmenides. Class Philol 40(4):223–242
11. Goldstein S (2015) The evolution of intelligence. In: Goldstein S, Princiotta D, Naglieri JA (eds) Handbook of intelligence: evolutionary theory, historical perspective, and current concepts. Springer, pp 3–7
12. Heller M (2012) Where physics meets metaphysics. In: Majid S (ed) On space and time. Cambridge University Press, pp 238–277
13. Israel DJ (1991) A short sketch of the life and career of john mccarthy. In: Lifschitz V (ed) Artificial intelligence and mathematical theory of computation: papers in honor of John McCarthy. Academic Press, Inc., pp 1–5
14. Kenny A (1994) Aquinas on mind. Routledge, London and New York
15. Lanz P (2000) The concept of intelligence in psychology and philosophy. In: Cruse H, Dean J, Ritter H (eds) Prerational intelligence: adaptive behavior and intelligent systems without symbols and logic, vol 1. Springer, pp 19–30
16. McCarthy J (1996) Defending AI research: a collection of essays and reviews. CSLI lecture notes no. 49, California

17. McClure J (2014) Conceptual parallels between philosophy of science and cognitive science: artificial intelligence, human intuition, and rationality. Aporia 24(1):39–49
18. Sarle JR (1983) Intentionality. An essay in philosophy of mind. Cambridge University Press, New York
19. Skansi S (2018) Introduction to deep learning: from logical calculus to artificial intelligence. Springer, Berlin
20. Sternberg RJ (1990) Metaphors of mind: conceptions of the nature of intelligence. Oxford University Press, New York
21. Sternberg RJ (2003) Wisdom, intelligence, and creativity synthesized. Cambridge University Press, Cambridge
22. Sternberg RJ, Berg CA (1986) Quantitative integration: definitions of intelligence: a comparison of the 1921 and 1986 symposia. In: Sternberg RJ, Detterman DK (eds) What is intelligence? Contemporary viewpoints on its nature and definitions. Ablex Publishing Corporation, pp 155–162
23. Turing AM (1950) Computing machinery and intelligence. Mind 59(236):433–460
24. Zagar J (1984) Acting on principles: a thomistic perspective in making moral decisions. University Press of America, Inc

Chapter 4
Why Not Fuzzy Logic?

Ivan Restović

Abstract Fuzzy logic is an approach to AI which focuses on the mechanization of natural language. It has long been proposed by Zadeh, its originator, as another paradigm for AI and the correct way to achieve "human level machine intelligence". To present day, this approach hasn't prevailed, but in the light of some recent tendencies in AI development it can gain traction. The "black-box property" of the currently predominant method—deep learning—has recently sparked a movement called "explainable artificial intelligence", a quest for AI that can explain its decisions in a way understandable and acceptable to humans. As it has been recognized, a natural way to provide explanations to users is to use natural language, embedded in the fuzzy logic paradigm. However, to model natural language fuzzy logic uses the notion of "partial truth", which has brought some philosophical concerns. The very core tenets of fuzzy logic have often been described as counterintuitive. In this text, we provide philosophical support for fuzzy logic by providing possible answers to the two most common critiques raised about it, as well as by offering independent philosophical motivation for endorsing it.

Keywords eXplainable Artificial Intelligence (XAI) · Natural language · Fuzzy logic · Philosophy of vagueness · Higher-order vagueness · Contradictions

4.1 A Problem and a Movement

Machine learning with deep neural networks is the predominant method in AI. However, there are some concerns about one of its widely recognized properties—many of the results of deep learning algorithms remain intransparent to humans. They reach a certain decision, but cannot provide reasons for it. This is usually called "the black-box property". Now, in certain areas it becomes "the black-box *problem*", e.g., in the field of medicine or in the financial sector.

I. Restović (✉)
Institute of Philosophy, Zagreb, Croatia
e-mail: ivan@ifzg.hr

S. Skansi (ed.), *Guide to Deep Learning Basics*,
https://doi.org/10.1007/978-3-030-37591-1_4

This has sparked a movement in AI in 2016[1] coined "eXplainable Artificial Intelligence" or XAI, proposed by Gunning on behalf of the USA Defense Advanced Research Projects Agency (DARPA). The present situation is described as such:

> The current generation of AI systems offer[s] tremendous benefits, but their effectiveness will be limited by the machine's inability to explain its decisions and actions to users. Explainable AI will be essential if users are to understand, appropriately trust, and effectively manage this incoming generation of artificially intelligent partners. [6, p. 2]

The proposal for XAI provides a summary of some existing AI techniques based on two features: performance versus explainability. Deep learning scores the highest on performance, but has very low explainability. Bayesian belief networks offer better explainability, but lag behind on performance. The best explainability is provided by decision trees, but there we also find the lowest performance. The desideratum is, of course, more explainability without loss in performance.

Although not mentioned by DARPA's proposal, many researchers have recognized the potential of fuzzy logic paradigm to assist XAI [1, 2, 7, 11]. As Alonso puts it, "interpretability is deeply rooted in the fundamentals of fuzzy logic" [1, p. 245]. This logic with its supporting theories and implementations has long been proposed as another paradigm for AI, most vigorously by its originator, Lotfi A. Zadeh. But is it itself understandable and acceptable to humans?

4.2 Zadeh's Proposal

Throughout his career, Zadeh argued for a paradigm shift in AI development. His position can be illustrated by this often paraphrased place:

> Humans have many remarkable capabilities; there are two that stand out in importance. First, the capability to reason, converse and make rational decisions in an environment of imprecision, uncertainty, incompleteness of information, partiality of truth and possibility. And second, the capability to perform a wide variety of physical and mental tasks without any measurements and any computations. A prerequisite to achievement of human level machine intelligence is mechanization of these capabilities and, in particular, *mechanization of natural language understanding*. In my view, mechanization of these capabilities is beyond the reach of the armamentarioum of AI – an armamentarioum which in large measure is based on classical, Aristotelian, bivalent logic and bivalent-logic-based probability theory. [22, p. 11, added emphasis]

Zadeh talks about the "achievement of human level machine intelligence". In the present context, we will slightly specify his claim. What we are looking for is "human *understandable* machine intelligence". Zadeh's original term may be misleading because it can be argued that some machines already surpassed the human level of intelligence. AI outperforms humans on a variety of tasks. There are forms of artificial intelligence *alien* to us—this makes the problem of XAI all the more urgent.

[1] According to [1, p. 244]. We use an updated version from 2017 [6].

In the context of transparent AI, one of the more substantial claims made in the above quote is one about natural language. Humans reason (mostly) in natural language. In Zadeh's opinion, this amounts to saying that we take as inputs sentences in natural language and after some "computation" output a conclusion, also in natural language. Now, wouldn't it be nice if computers reasoned in natural language in the way humans do? Zadeh's working assumptions on natural language can be seen here:

> Much of human knowledge is expressed in natural language. [...] The problem is that natural languages are intrinsically imprecise. Imprecision of natural languages is rooted in imprecision of perceptions. A natural language is basically a system for describing perceptions. Perceptions are intrinsically imprecise, reflecting the bounded ability of human sensory organs, and ultimately the brain, to resolve detail and store information. Imprecision of perceptions is passed on to natural languages. [21, p. 2769]

This passage seems to imply that natural language is enough to store human knowledge based on perception. In other words, that there is nothing in perceptions which cannot be expressed in the natural language. This is clearly an even more substantial and also a controversial claim. However, in the context of XAI, we don't need to fully endorse it. Maybe there is something "lost in translation" but it is lost on both sides since the question posed by a human is itself in natural language. This text is only about "linguistic explanations" given by a logical system designed to resemble human reasoning.

Zadeh proposed several closely connected theories for implementing his above-described motivation, some of which we mention here. For modeling the underlying perceptions, he proposes the Computational Theory of Perception (CTP) wherein perceptions and queries are expressed as propositions in natural language. Having perceptions thus modeled, we can use CTP's underlying methodology of Computing With Words (CWW) to yield answers to queries [20]. Computing With Words in turn is a branch of fuzzy logic in the broad sense [19], but it is also based on fuzzy logic in the narrow sense, a logic of approximate reasoning [17]. More on this ambiguity shortly.

Fuzzy logic employs a nonclassical set of truth values: they are considered as belonging to the unit interval [0, 1], in accordance with the notion of fuzzy sets Zadeh introduced in [15]. The basic notion of this set theory is *partial elementhood.* In fuzzy logic, there is partial elementhood in the "set" of truth values.

Partiality of truth was introduced to capture the intuition that for some concepts there are no clear boundaries. In [15] Zadeh wonders if bacteria are animals. The answer might be—partly. In fuzzy logic atomic propositions are assigned truth values in the interval [0, 1]. Truth conditions for connectives are taken from Łukasiewicz[2]:

$$v(\neg p) =_{def} 1 - v(p)$$
$$v(p \wedge q) =_{def} min(v(p), v(q))$$
$$v(p \vee q) =_{def} max(v(p), v(q))$$
$$v(p \rightarrow q) =_{def} min(1, 1 - v(p) + v(q)).$$

[2] Appearing in his paper from 1930. For an English translation see [9].

What brings fuzzy logic closer to natural language is its use of *linguistic variables* [16] for truth values. Even though there is an underlying computation, we wouldn't get an answer like "*Bacteria are animals* is 0.892 true" since it would be far from natural language. In [17, p. 410] Zadeh uses the countable set {*true, false, not true, very true, not very true, more or less true, rather true, not very true and not very false, ...* }. For instance, we can label as "true" those propositions the value of which exceeds 0.5. The threshold for "very true" can be 0.7, and so on.

"Fuzzy logic" can mean different things. In the broad sense, it includes all the theories Zadeh proposes for mechanization of natural language, some of which are more specialized than the others. So, when Zadeh opts for a paradigm shift toward fuzzy *logic*, he doesn't mean that the whole work in a field as broad as AI has to be done solely within logic as a subfield of pure mathematics or of philosophy. Then, in the narrow sense, "fuzzy logic" signifies such a subfield, i.e., the logical system with truth values in the unit interval and with linguistic variables for such values.

Because of its focus on (computing with) natural language, fuzzy logic has been recognized as a viable approach to XAI [1, 7, 11]. Even if Zadeh's insistence on the use of natural language was exaggerated, this feature is now extremely useful for giving *logical* explanations acceptable to humans.

For instance, Hagras states:

> [...] FRBS [fuzzy rule-based system] generates if-then rules using linguistic labels (which can better handle the uncertainty in information). So, for example, when a bank reviews a lending application, a rule might be: if income is *high* and *home owner* and *time* in *address* is *high*, then the application is deemed to be from a good customer. Such rules can be read by any user or analyst. More importantly, such rules get the data to speak the same language as humans. [7, p. 35]

We will not debate the understandability of Hagras' example rule. Notice just that italicized words represent variables, some of which are fuzzy terms. Consider "high income". It would not be useful for a bank to classify incomes only according to two categories: high versus low. Two incomes *a* and *b* can both be low, but one can still be higher than the other. In fuzzy logic this amounts to saying that the sentence "Income *a* is high" is more true than the sentence "Income *b* is high". Similarly, Alonso [1] sees Zadeh's CWW especially relevant to XAI since humans are used to explanations in natural language.

However, a host of philosophical critiques are raised against fuzzy logic. A great deal of them attacks even its fundamental tenets, like the very notion of partial truth. In the following section, we provide philosophical support for fuzzy logic. First, we analyze the philosophical setting in which fuzzy logic is often proposed—the sorites paradox. Then we describe the two common concerns raised about the viability of fuzzy logic and outline possible answers. The last part offers an intermediate position, tenable even if some critiques against fuzzy-set-theoretic treatment of truth values are left unanswered.

4.3 Philosophical Concerns

4.3.1 *Fuzzy Logic and the Sorites Paradox*

In philosophy, fuzzy logic is often considered as a special solution for a more general problem—vagueness, or the possibility of a concept to have borderline cases. Vagueness is problematic because it invites the famous sorites (heap) paradox.

Let's illustrate this by using the most popular predicate in the literature about fuzzy logic, "tall". Consider Sandy Allen, the American actress who was 231 cm tall. Now, everyone would agree that the proposition "The person standing at 231 cm is tall". Also, it seems plausible to affirm the conditional: "If a person standing at x cm is tall, so is the person standing at x cm − 1 mm". In other words, if there was a person only 1 mm shorter than Allen, they would still be considered tall; a tenth of a centimeter doesn't make a difference.

But if we were to line up actors and actresses by their height starting with Sandy Allen and apply the conditional a number of times, we would get counterintuitive results. For instance, it would follow that Danny DeVito, standing at 147 cm, is tall. This is clearly not the right result, despite the premises and the rules of inference being acceptable. How to make DeVito short again? Similarly, we can start with DeVito and the conditional "If a person standing at x cm is short, so is the person standing at x cm + 1 mm", which would in turn make Allen short.

Introducing partiality of truth can help us solve the paradox. Allen is clearly tall. DeVito is clearly not. Fuzzy logic can get us to the right conclusion. Just as the height of people in our lineup decreases, so do the truth values of height ascriptions to people down the sorites. Also, the conditionals are all almost fully true. There is nothing paradoxical about sorites in the fuzzy logic paradigm. The paradox appears only when we use bivalent definitions for fuzzy concepts [22]. Tall is a fuzzy concept, and it should be modeled as such.

Consider again our example. To make comparisons one-dimensional, we will only speak about heights of actresses. Now, we all agree that:

v(Sandy Allen (231 cm) is tall) = 1 (i)

But obviously, she is not the only one who deserves to be classified as "fully tall", i.e., for whom the truth value of height ascription proposition is 1.[3] Let's decide that the last actress to be fully tall is Geena Davis:

v(Geena Davis (183 cm) is tall) = 1 (ii)

So, anybody shorter than her would have $v < 1$ as a truth value of the proposition ascribing height.

On the other end of the spectrum we have some actresses that are clearly not tall. Call this "short". We now have to decide on the tallest "fully not tall" or the tallest "absolutely short" actress on the list. We decide on:

[3]Firstly, there could be an even taller actress. Since we cannot give a value above 1, she would get the same truth value as Allen. Note that, in the present case, it would not make sense for the arrival of a newcomer to lower the truth value of the initial "truest" element.

v(Judy Garland (151 cm) is tall) $= 0$ (iii)
So, anybody taller than her would have $v > 0$ as a truth value of the proposition ascribing height.

Let's now calculate the intermediate case. It is someone of a height that is in between the two cutoff points. We will characterize it thus:
v("Judy Davis" (167 cm) is tall) $= 0.5$ (iv)
This would put Meryl Streep (168 cm) slightly on the taller side. Let's approximate:
v(Meryl Streep (168 cm) is tall) $= 0.55$ (v)
Assume that there is no one between "Davis" and Streep. In that case, the latter actress is the last person in our sorites series to whom tallness may be ascribed more than shortness. So, not everybody is tall—we don't get the counterintuitive result as in the case of bivalent logic.

The presented scale looks useful, but as the reader might have noticed, we have made some questionable assumptions. We had to draw two borders, both of which seem arbitrary. And that in turn resulted in an also seemingly arbitrary intermediate case between the two. Had Geena Davis not been absolutely tall, Meryl Streep might have ended on the other side of the boundary.

4.3.2 *The Problem of Higher-Order Vagueness*

What we have encountered is the problem of higher-order vagueness or arbitrary precision (cf. [14, Chap. 4], [8, Chaps. 4–5]). As Keefe puts if, "what could determine which is the correct function, settling that my coat is red to degree 0.322 rather than 0.321?" [8, p. 114]. She argues that a function from measurements to truth values for every fuzzy concept should be unique. Otherwise, we lose the ordering relation between sentences which was supposed to be an asset of fuzzy logic. If the same coat is also blue to the degree 0.321, is it now more red than blue? However, this uniqueness is unwarranted since there is no clear-cut answer on how to acquire the initial truth values.

Similarly, Williamson argues that in fact, sentence like (v) above are vague rather than exact. So, although truth by numbers looks like a more precise and nuanced account than the classical picture, it doesn't resolve the original problem. Is the sentence (v) *absolutely* true? "Even if statistical surveys of native speaker judgements were relevant to deciding [...], the results would be vague. It would often be unclear whom to include in the survey, and how to classify the responses" [14, p. 128].

Both Keefe and Williamson propose different accounts of truth value ascription to vague statements. Williamson proposes a position now called "epistemicism". Vagueness is just ignorance, there is nothing vague or fuzzy about the world. Every sentence is either true or false. Even the seemingly borderline sentence (iv). On that view, making the notion of truth more nuanced doesn't help our lack of knowledge, as it is shown by the problem of higher-order vagueness [14, Chaps. 7–8]. Keefe, on the other hand, argues for "supervaluationism". On that view, there are nonclassical values, but these values are not truth-functional. Some sentences fall in a "truth-value gap" [8, Chaps. 7–8].

We will now outline two possible ways of alleviating the problem of higher-order vagueness for a proponent of fuzzy logic. First one more philosophical, the other more mathematical.

Take Keefe's question about finding out the exact value of redness for her coat. Smith [12] argues that this is not the job for fuzzy logic. Fuzzy logic is a calculus of fuzzy truth values. Given such values for atomic propositions, we use logical laws to infer other truths. What these truths are is a matter for another discipline:

> Classical logic countenances only two truth values [...]. This does *not* make it correct, however, to say that it is a commitment of classical logic (model theory) that every statement is either true or false. Such a commitment comes into play only when one seeks to use classical logic to shed light on the semantics of some language (e.g., natural language, or the language of mathematics). It is thus a commitment not of pure classical logic (model theory) – considered as a branch of mathematics – but of model-theoretic semantics (MTS). [...] Pure model theory tells us only that a wff is true on this model and false on that one (etc.). [12, p. 2]

Of course, the problem is here not solved, just relocated. Smith [12] recognizes that and offers possible answers. But from the standpoint of pure logic, one can lessen the concern by endorsing a working assumption of logical pluralism: different logics can be used for different purposes. Remember intuitionism in mathematics. Brouwer [4] argued that real mathematics doesn't conform to some laws of classical logic, most famously the principle of excluded middle. But in other domains, such as reasoning about our everyday finite domains, there is no fault in using classical logic. However, it took a different kind of research to come to the true nature of mathematics and its corresponding logic. The research into the correct fuzzy truth values may turn out to be in the scope of fuzzy logic in the broad sense, but this shouldn't hinder the progress of fuzzy logic in the narrow sense.

Especially since if truth *simpliciter* is a logical or mathematical problem, it is not so just for fuzzy logic. Even classical predicate logic cannot decide on a truth value of the proposition "Bacteria are animals". It can only say what follows from that proposition. Every logic is about valid reasoning, arriving at true conclusions *given* true premises, which often come from other areas of knowledge. Fuzzy logic, we argue, can be the right way to describe some phenomena.

The mathematical way to combat higher-order vagueness is to admit that in some cases the truth value of a proposition is not unique, but that this can be accounted for set-theoretically. Along with "regular" fuzzy sets, Zadeh [16] proposed fuzzy sets with fuzzy membership functions. In that way, we can model second order, as well as higher-order vagueness.

> A fuzzy set is of type n, $n = 2, 3, \ldots$, if its membership function ranges over fuzzy sets of type $n - 1$. The membership function of a fuzzy set of type 1 ranges over the interval [0, 1]. [16, p. 242]

Let's illustrate this within our example. The average height for an actress was 167 cm. But there seem to be other appropriate ascriptions. Let's say we have several authorities on height, who don't all agree. The lowest proposed average height is 162 cm, and the largest 170. Now we can fuzzify the concept of average height. It is not exactly 167 cm, but somewhere in between—it becomes an interval, rather than a point. There is a "footprint of uncertainty" [7, p. 34] on the scale. Note that the actresses over 170 cm tall are still clearly above average.

All this being said, one might still claim that some terms like "beautiful" seem to resist mathematical treatment. Height is easy to capture since there is only one variable to measure. But how to find out which variables to consider in the correct "beauty-function"? Here it would be useful to introduce a notion of a "prototype". Something may be said to be beautiful to such-and-such degree of truth depending on its closeness to some prototype(s). In the area of psychology of concepts, this is a well-known approach and some of the groundbreaking work was influenced by Zadeh himself. See [3] for a discussion about fuzzy logic in this area.

4.3.3 The Problem with Contradictions

Putting aside the problem with arriving at the initial truth values, there is yet another concern often raised against fuzzy logic, one that actually *is* in its providence—it allows for true contradictions. As stated above, in fuzzy logic, the truth value of $\neg p$ ($v(\neg p)$) is defined as $1 - v(p)$. Also, conjunction assumes the same value as the lowest conjunct (*min* function).

Previously, we defined "short" as the negation of "tall". Considering the clear cases of tallness, we can assert (see proposition (ii)):

v(Geena Davis is tall and short) $= 0$ (vi)

Davis is fully tall and not at all short. The conjunction takes the lesser value and turns out totally false, just like in classical logic. However, the problem appears among intermediate cases. For we have:

v("Judy Davis" is tall and short) $= 0.5$ (vii)

v(Meryl Streep is tall and short) $= 0.45$ (viii)

Numerous authors have criticized fuzzy logic for this feature. So much so that Smith calls it "the undead argument" [13]. He outlines several lines of response to this argument, coming from several disciplines.

Philosophers usually label the sentences (vii–viii) as *counterintuitive*: the principle of non-contradiction is the undisputed logical axiom and should (fully) hold in all theories. However, this may be circular. Fuzzy logic is accused of not following the classical principles. But it is exactly the inability of classical logic to model the "real world" that was the motivation for a nonclassical approach, such as fuzzy logic. Zadeh simply has different intuitions about bivalence, as we saw from his proposal for a paradigm shift.

Also, note that there are no blatant contradictions in fuzzy logic. Contradictions can be at most half-true. Nothing is both a triangle and a circle, both in classical and in fuzzy logic. This is because such concepts are not vague. Not-absolutely-false contradictions appear only with vague concepts, which classical logic cannot model in the first place.

Returning to different intuitions, consider again the putatively controversial proposition (viii). From it we can infer:
Meryl Streep is more tall than short. (ix)
We don't consider this proposition neither blatantly false nor meaningless, even if it rests on a contradiction. Streep *is* both tall and short but she is also more tall than short, which can be seen as just another way of saying that her height is slightly above average.

The situation seems to be even more clear in the case of our "most true contradiction", proposition (vii). Asserting it amounts to saying:
"Judy Davis" is as tall as she is short. (vii')
Again, we don't see anything wrong with this assertion. Our hypothetical actress is right in the middle, and a contradiction of a value 0.5 tells us exactly that. So, one could instead argue that, contrary to being unintelligible, there is additional information in true contradictions in fuzzy logic. Whereas in classical logic they all get the same truth value, in fuzzy logic their truth value tells us more [3, p. 31]. This logic is simply more expressive than its classical counterpart.

4.3.4 Vagueness Is Not Fuzziness

In the preceding text, we have treated fuzzy logic based on fuzzy set theory as an answer to the problem of vagueness. This view has been prevalent in the philosophical literature. However, fuzziness can be seen as distinct from vagueness. Importantly, this is the view expressed by Zadeh himself. Dubois [5] further elaborates and expands on this view expressed in the following quote, showing that both epistemicism (vagueness as ignorance) and supervaluationism (there is a gap in truth value for borderline cases) are compatible with the notion of truth modeled by fuzzy sets. Zadeh argues:

> Although the terms *fuzzy* and *vague* are frequently used interchangeably in the literature, there is, in fact, a significant difference between them. Specifically, a proposition, p, is *fuzzy* if it contains words which are labels of fuzzy sets; and p is *vague* if it is both fuzzy and insufficiently specific for a particular purpose. For example, "Bob will be back in a few minutes" is fuzzy, while "Bob will be back sometime" is vague if it is insufficiently informative as a basis for a decision. Thus, the vagueness of a proposition is a decision-dependent characteristic whereas its fuzziness is not. [18, p. 396, n.]

Here we see that vagueness includes fuzziness, but there is another important characteristic of vague sentences—they don't offer enough information to be accounted for by fuzzy sets. With this distinction at hand, we can accommodate some theories about vague propositions. One can claim that there *are* truth value gaps, but they only concern vague propositions. Such propositions are in a way deficient, they are too underspecified to be ascribed a numerical truth value, be it classical of fuzzy, even type-n fuzzy. On the other hand, there is nothing underspecified in an exclusively fuzzy description of a predicate. In Dubois' words: "While vagueness is a defect, gradualness is an enrichment of the Boolean representation" [5, p. 317].

Epistemicist theory of vagueness can also be incorporated to fit this distinction. Vagueness is still ignorance, not of just two possible truth values, but of the exact fuzzy truth value. Dubois calls this a "gradual epistemic view", according to which partially true propositions exist, but they appear vague or imprecise because of our (partial) ignorance. Similar point is elaborated by MacFarlane [10] in a view called "fuzzy epistemicism". Classical (bivalent) epistemicism claims that what distinguishes vague language from non-vague language has only to do with our knowledge, not with the underlying metaphysics of truth. However, "both uncertainty and partial truth are needed to understand our attitudes towards vague propositions" [10, p. 438].

This concerns some cases of higher-order vagueness. Firstly, if fuzzy epistemicism is correct, some first-order vagueness can actually be downgraded to fuzziness via amelioration of our epistemic position. And if there is still some vagueness about such fuzziness, it can again be a result of insufficient specificity. If so, it can then be alleviated with the corresponding type-n fuzzy sets. It may take some conceptual analysis to come to know the "depth" of a (putatively) vague concept or proposition, but once we find that level, we can describe it mathematically.

4.4 Conclusion

Machine learning with deep neural networks is the prevailing paradigm of AI. However, the black-box property of deep learning algorithms may often propose a problem. This has recently sparked a movement called eXplainable Artificial Intelligence (XAI). Decisions made by AI should seek to become more transparent to humans.

Now, humans are the most accustomed to explanations in natural language. And it is exactly the insistence on natural language that is the hallmark of another approach to AI, Zadeh's fuzzy logic paradigm, which has been recognized as a viable approach toward XAI. This paradigm rests on "fuzzy logic" in the narrow sense, i.e., a logical calculus of partial truth.

However, it has been argued that fuzzy logic is not meaningful or acceptable (to humans) since some of its fundamental notions are mistaken or unintelligible. The aim of this text was to provide philosophical support for fuzzy logic. We first described the most common philosophical motivation for introducing this nonclassical logic—the sorites paradox. Then we addressed two common critiques. Fuzzy logic has been accused of harboring higher-order vagueness and allowing for true contradictions.

It is argued that such a nuanced view of a truth value as a number in the interval [0, 1] is itself vague since there is no transparent way of finding the exact value. We proposed two ways of alleviating higher-order vagueness. Firstly, it can be argued that finding the right (fuzzy) truth values for atomic propositions is not the domain of (fuzzy) logic. Secondly, even if in some cases the numbers are not unique, fuzzy set theory can be expanded via type-n fuzzy sets to mathematically describe this phenomenon.

Connectives in fuzzy logic are so defined as to allow some contradictions not to be fully false. This is often considered an undesirable feature which any correct theory should avoid. However, it is important to note that in fuzzy logic contradictions are at most half-true. We explored some true contradictions and argued that they can indeed be meaningful and even informative.

We also proposed arguments for distinguishing vagueness from fuzziness. In philosophy, fuzzy logic is often seen as just another theory of vagueness along with competing theories such as "epistemicism" and "supervaluationism". However, it can be argued that vagueness includes both fuzziness and an additional characteristic—lack of information. On this view, fuzzy logic doesn't compete with theories of vagueness—they can work in concert.

The notion of partial truth turns out not to be as counterintuitive as it first appears. This being the case, we think it is safe to assume that explanations provided by AI arrived at by using fuzzy logic can be understandable to humans, especially provided an accessible and coherent underlying philosophy of fuzziness.

References

1. Alonso JM (2019) From Zadeh's computing with words towards eXplainable artificial intelligence. In: Fullér R, Giove S, Masulli F (eds) Fuzzy logic and applications. Springer, Cham, pp 244–248
2. Alonso JM, Castiello C, Mencar C (2018) A Bibliometric analysis of the explainable artificial intelligence research field. In: Medina J, Ojeda-Aciego M, Verdegay JL, Pelta DA, Cabrera IP, Bouchon-Meunier B, Yager RR (eds) Information processing and management of uncertainty in knowledge-based systems. Theory and foundations (Part I). Springer, Cham, pp 3–15
3. Belohlavek R, Klir GJ, Lewis III HW, Way EC (2009) Concepts and fuzzy sets: misunderstandings, misconceptions, and oversights. Int J Approx Reason 51(1):23–34
4. Brouwer LEJ (1949) Consciousness, philosophy and mathematics. In: Beth EW, Pos HJ, Hollak JHA (eds) Proceedings of the tenth international congress of philosophy. North-Holland Publishing Company, Amsterdam, pp 1235–1249
5. Dubois D (2011) Have fuzzy sets anything to do with vagueness? In: Cintula P, Fermüller CG, Godo L, Hájek P (eds) Understanding vagueness. Logical, philosophical and linguistic perspectives (Studies in logic). College Publications, London, pp 311–333
6. Gunning D (2017) Explainable artificial intelligence (XAI). Def Adv Res Proj Agency (DARPA) 2:1–36. https://www.darpa.mil/attachments/XAIProgramUpdate.pdf. Accessed 15 Aug 2019
7. Hagras H (2018) Toward human-understandable, explainable AI. Computer 51(9):28–36
8. Keefe R (2000) Theories of vagueness. Cambridge University Press, Cambridge
9. Łukasiewicz J (1970) Investigations into the sentential calculus. In: Borkowski L (ed) Jan Łukasiewicz: selected works. North-Holland Publishing Company, Amsterdam and London, pp 131–152
10. MacFarlane J (2009) Fuzzy epistemicism. In: Dietz R, Moruzzi S (eds) Cuts and clouds. Vagueness, its nature, and its logic. Oxford University Press, New York, pp 438–463
11. Mencar C (2019) Looking at the branches and roots. In: Fullér R, Giove S, Masulli F (eds) Fuzzy logic and applications. Springer, Cham, pp 249–252

12. Smith NJJ (2011) Fuzzy logic and higher-order vagueness. In: Cintula P, Fermüller CG, Godo L, Hájek P (eds) Understanding vagueness. Logical, philosophical and linguistic perspectives (Studies in logic). College Publications, London, pp 1–19
13. Smith NJJ (2017) Undead argument: the truth-functionality objection to fuzzy theories of vagueness. Synthese 194(10):3761–3787
14. Williamson T (1994) Vagueness. Routledge, London
15. Zadeh LA (1965) Fuzzy sets. Inf Control 8(3):338–353
16. Zadeh LA (1975) The concept of a linguistic variable and its application to approximate reasoning—I. Inf Sci 8(3):199–249
17. Zadeh LA (1975) Fuzzy logic and approximate reasoning. Synthese 30(3–4):407–428
18. Zadeh LA (1978) PRUF—a meaning representation language for natural languages. Int J Man-Mach Stud 10(4):395–460
19. Zadeh LA (1996) Fuzzy logic = computing with words. IEEE Trans Fuzzy Syst 4(2):103–111
20. Zadeh LA (1999) From computing with numbers to computing with words—from manipulation of measurements to manipulation of perceptions. IEEE Trans Circuits Syst I: Fundam Theory Appl 46(1):105–119
21. Zadeh LA (2008) Is there a need for fuzzy logic? Inf Sci 178(13):2751–2779
22. Zadeh LA (2008) Toward human level machine intelligence—is it achievable? The need for a paradigm shift. IEEE Comput Intell Mag 3(3):11–22

Chapter 5
Meaning as Use: From Wittgenstein to Google's Word2vec

Ines Skelac and Andrej Jandrić

Abstract Modern natural language processing (NLP) systems are based on neural networks that learn concept representation directly from data. In such systems, concepts are represented by real number vectors, with the background idea that mapping words into vectors should take into account the context of their use. The idea is present in Wittgenstein's both early and late works, as well as in contemporary general linguistics, especially in the works of Firth. In this article, we investigate the relevance of Wittgenstein's and Firth's ideas for the development of Word2vec, a word vector representation used in a machine translation model developed by Google. We argue that one of the chief differences between Wittgenstein's and Firth's approaches to the word meaning, compared to the one applied in Word2vec, lies in the fact that, although all of them emphasise the importance of context, its scope is differently understood.

Keywords Wittgenstein · Firth · Word2vec · Machine translation · Natural language processing

5.1 Introduction

Meaning has always been one of the essential topics in philosophy since its origin is in Ancient Greece. Some of the important philosophical questions related to meaning are: what is the relation between language and thought? what is a concept? is it a mental image? how do elements of language refer to non-linguistic entities? how are we able to know the meaning of a word or a sentence?, etc.

I. Skelac (✉)
Algebra University College, Zagreb, Croatia
e-mail: ines.skelac@racunarstvo.hr

A. Jandrić
Faculty of Philosophy, University of Belgrade, Belgrade, Serbia
e-mail: ajandric@f.bg.ac.rs

S. Skansi (ed.), *Guide to Deep Learning Basics*,
https://doi.org/10.1007/978-3-030-37591-1_5

On the other hand, in recent developments of artificial intelligence (AI) meaning has proved to be one of the greatest challenges. So far, artificial intelligence has been made skilled in, among other things, recognising speech and translating text and/or speech from one language to another, but it is still short of understanding human language. The problem of natural language understanding cannot yet be solved by using AI technology alone, and until now it has required huge manual efforts. However, an important step forward has been made with introducing neural networks in natural language processing, especially in machine translation.

Although it is still an active field of research, modern natural language processing (NLP) systems are based on neural networks that learn concept representation directly from data, without human intervention. In such systems, concepts are represented by vectors of real numbers. The idea which serves as the basis for models of word embedding using neural networks is that mapping words into vectors should take into account the context of a sentence because its meaning is not a simple composition of the meanings of individual words it contains. To learn vector representation for phrases, it is necessary to find words that appear frequently together, and infrequently in other contexts [7].

This fundamental idea sounds almost like a repetition of Frege's famous dictum: "Never ask for the meaning of a word in isolation, but only in the context of a sentence" ([4]: x). The idea is present in Wittgenstein's both early and later works. In the *Tractatus*, he says that "Only the proposition has sense; only in the context of a proposition has a name meaning" ([11]: Sect. 3.3), while in *Philosophical Investigations* he develops it further: "We may say: nothing has so far been done, when a thing has been named. It has not even got a name except in the language game. This was what Frege meant too when he said that a word had meaning only as part of a sentence" ([12]: Sect. 49). Outside philosophy, in contemporary general linguistics, the idea of the importance of context for establishing meaning became relevant through the works of J. R. Firth: for Firth, the complete meaning of a word is always contextual ([3]: 7).

In this article, we investigate the relevance of Wittgenstein's and Firth's ideas, as referred to above, for the development of word vector representation used in a machine translation model developed by Google, known as Word2vec. In Word2vec, the meaning of a word is identified with a vector (in standardised form) which codifies the contexts of its use; a particular context in which the word occurs is characterised as the words immediately surrounding it in a sentence. We will first present Wittgenstein's and Firth's views on the importance of context for determining word meaning. After that, we will briefly outline the most important technological aspects of Word2vec. Finally, we will compare these approaches, and highlight similarities and differences between them.

5.2 The Role of Context in Wittgenstein's Philosophy of Language

Arguably, no twentieth-century philosopher stressed so vehemently the importance of context in determining word meaning, and thereby radically transformed our understanding of the workings of language, as Ludwig Wittgenstein.

Although it is widely recognised that context plays a prominent role in Wittgenstein's later philosophy of language, for reasons of historical accuracy it should be underscored that Wittgenstein, from the very beginning, ascribed high importance to the so-called context principle, according to which words have no meaning in isolation. Already in the *Tractatus*, it is stated in Sect. 3.3: "Only the proposition [*Satz*] has sense [*Sinn*]; only in the context of a proposition [*Zusammenhange des Satzes*] has a name meaning [*Bedeutung*]. "Wittgenstein derived the context principle, as well as the terminology in which it is expressed, through the influence of "the great works of Frege", to which he admits owing "in large measure the stimulation of my thought" ([11]: Preface).

In his *Grundlagen der Arithmetik*, Gottlob Frege mentions the context principle twice: in the foreword, he stresses that one of his fundamental principles is "never to ask for the meaning of a word in isolation, but only in the context of a proposition [*Satzzusammenhange*]" ([4]: x); and later, in Sect. 60, he advises that "we ought always to keep before our eyes a complete proposition [*Satz*]", since "only in a proposition have the words really a meaning [*Bedeutung*]". For Frege, the context principle is mainly a powerful shield against the temptation to succumb to psychologism in the philosophy of mathematics. Since the numerical singular terms do not stand for physical objects that we find in our experience, in attempting to determine their reference (*Bedeutung*) without considering the sentential context in which they occur we are often inclined to wrongly assume that they stand for mental objects, or ideas (*Vorstellungen*), and that, accordingly, a psychological investigation of these ideas will provide us with the foundations of mathematics. Another reason which Frege had for endorsing the context principle is to be found in his subsequent article "On Concept and Object": there he points out that a word can refer to entities of radically different kinds in various sentences, and that, hence, its reference cannot be determined outside a specific context. For instance, in the sentence "Vienna is a big city" the word "Vienna" behaves like a name and thus stands for an object, a saturated entity, namely, the capital of Austria; but in the sentence "Trieste is no Vienna" it has the role of a predicate and refers to an unsaturated entity, a concept, namely, that of being a metropolis ([5]: 50).[1]

---

[1]Frege divided linguistic expressions into saturated (or names) and unsaturated (or functional) expressions: the unsaturated expressions contain a gap, an empty place for an argument usually marked by the occurrence of a variable. This linguistic division is strictly paralleled by an ontological division: references of saturated expressions are saturated entities (or objects), while references of unsaturated expressions are unsaturated entities (or functions). Predicates are functional expressions which need to be supplemented with a name to form a proposition: analogously, at the level of reference, concepts, which are references of predicates, are first-order functions of one argument,

Wittgenstein opposed psychologism in the philosophy of mathematics as ardently as Frege. In fact, he considered psychological investigations of mental processes accompanying the uses of language as irrelevant for explaining meaning, not only of mathematical terms but also of words in general: his early statements against such a view can already be found in the *Tractatus* ([11]: Sect. 4.1121), but their most elaborate and conclusive form was to wait until *Philosophical Investigations* ([12]: Sects. 143–184). The second reason Frege had for endorsing the context principle was also well received by Wittgenstein in the *Tractatus*: words may have different meanings in different sentences, which is why he warns us that we should distinguish between a mere sign (*Zeichen*) and the symbol (*Symbol*) it expresses, since "two different symbols can /.../ have the sign (the written sign or the sound sign) in common" ([11]: Sect. 3.321), as well as that "in order to recognise the symbol in the sign we must consider the significant use" ([11]: Sect. 3.326), that is, the use that the sign has been put to in the context of a meaningful sentence.

In the so-called transitional period, in the late 1920s, Wittgenstein's views on the context of use of a sign have undergone an important change.[2] He no longer believed that a sentence was the least independent linguistic unit endowed with meaning, but embraced a more large-scale semantic holism instead. According to his new understanding, a word has meaning only within a *Satzsystem*, a system of propositions relatively independent of the rest of the language. While in the *Tractatus* language was described as a monolithic great "mirror of the world" ([11]: Sect. 5.511), which represents the world by sharing its logical structure, in his new view the language was broken up into smaller, autonomous, overlapping linguistic systems, each constituted by its own set of explicit rules that prescribe the use of its primitive terms. A *Satzsystem* is, perhaps, best thought of by analogy to an axiomatic system. His motivation behind this change was twofold. One reason was his dissatisfaction with both the Fregean realism and formalist antirealism in the philosophy of mathematics: while formalists denied meaning to mathematical signs, realists claimed they referred to objects whose existence was completely independent of our thought, language and mathematical practice. Wittgenstein saw a fruitful third way in comparing mathematics to a game of chess: it is not true that chess figures have no meaning, only that they mean something exclusively within the game; their meaning is not to be identified with an object they refer to, but with rules of the game which command the moves one can make with them ([15]: 142–161). The other reason stemmed from his dissatisfaction with his own views expressed in the *Tractatus*. Discussing his book with Frank Ramsey made him realise that he had a problem with explaining how the proposition that something is red implies the proposition that it is not green. Early on, Wittgenstein espoused the view that has recently become known as modal monism: he believed that there is only one kind of necessity, namely, logical necessity ([11]: Sect. 6.37). While it is necessary that a red thing is not green at the same time, the

---

whose value is a truth value. Frege [5] claimed that the division is absolute: nothing can be an object in one context and a function in some other.

[2]A nice overview of the development of semantic holism in Wittgenstein can be found in Shanker [10].

necessity in question seems not to be logical: in order to reduce it to logical necessity, Wittgenstein was forced to claim that impossibility of anything being both red and green stems from the meaning of the colour words. In his new opinion, words such as "red" or "green" have meaning only inside the propositional system for attributing colours to objects, and the rules which constitute this system forbid predicating two colours to the same object ([14]: Sects. 76–86).

In the next phase of Wittgenstein's philosophical development, *Satzsysteme* were replaced with an even richer concept of *Sprachspiele*, or language games. Wittgenstein introduced the notion of a language game in the *Blue Book*, while in the *Brown Book* he gave many examples of them, some of which later reappeared in *Philosophical Investigations*. He describes language games as primitive forms of language, complete in themselves ([13]: 81), but easily imagined as evolving, in changed circumstances, into new and more complex ones ([13]: 17). In his many remarks, Wittgenstein suggests that it may be useful to think of language games as a language of a primitive tribe that one encounters ([13]: 81), since they are "ways of using signs simpler than those in which we use the signs of our highly complicated everyday life" ([13]: 17). He also compares them to "the forms of language with which a child begins to make use of words" ([13]: 17), and indicates that later in life one is initiated into novel language games when one, for instance, learns "special technical languages, e.g., the use of charts and diagrams, descriptive geometry, chemical symbolism, etc." ([13]: 81). Wittgenstein now sees ordinary language as a complicated network of interconnected language games, in which words are being used as extremely diverse tools for multifarious purposes ([12]: Sect. 11). The most important difference between *Satzsysteme*, which he previously considered as basic contexts of word use, and the subsequent *Sprachspiele* is that in language games the meaning of words is inextricably tied to speakers' non-linguistic practice: by the term "language game" Wittgenstein understands "the whole, consisting of language and the actions into which it is woven" ([12]: Sect. 7). To explain the words meaning in a language game, it is not sufficient, as in a *Satzsystem*, to lay down semantic rules, but it also needs to be specified who constitutes the linguistic community, what kind of non-verbal activities members of the linguistic community are typically engaged in when uttering the words, what props are thereby being used, what appropriate non-verbal reactions to hearing the words uttered are, how they are being taught to novices, and what customs and institutions already have to be in place so that the linguistic training may succeed and language application can get off the ground ([12]: Sects. 2–7). Wittgenstein repeatedly stresses that the meaningful use of language presupposes participation in a community, whose members must already agree in their behavioural responses, both to one another and to their common surroundings. Words have meaning only inside a language game ([12]: Sect. 49); their meaning is the way they are used therein ([12]: Sect. 43); since language games are already independent miniature languages, to understand a single word means to understand a whole language ([12]: Sect. 199); and as our linguistic behaviour can appear only against the background of shared and rule-regulated non-linguistic practices, mastering the technique of using words presupposes being initiated into a certain culture or a form of life ([12]: Sects. 19, 23, 199, 241).

When we use words in a proper environment, while taking part in activities in which these words are at home, misunderstandings only seldom occur and are easily resolved. On the other hand, we run into troubles, are confused and perplexed with paradoxes when we divorce words from their original surroundings: according to Wittgenstein, misuse of language is the source of all philosophical puzzles; they arise when "language goes on holiday" ([12]: 38). Philosophical problems should, in his view, disappear once the words are brought back to their everyday use: the successful treatment thus consists in producing an *Übersicht*,[3] a perspicuous presentation with which to remind ourselves of the roles that the words have in various contexts ([12]: Sects. 122–133). A particular source of philosophical troubles is that quite often words are used differently in different language games: if we successfully apply a word on a certain occasion, we are inclined to think that in a new context, in which its meaning has been altered, it must conform to the same rules as before. Another powerful philosophical prejudice that Wittgenstein was persistently striving to free us from is that there has to be a common core to all the different context-relative meanings of a word, a set of context-transcending necessary and sufficient conditions for its application. He pointed out that this is not the case with all words and that some words, such as "game", stand for family resemblance concepts: different cases of their use show similarities in pairs, even though there is no "one fibre running through the whole thread" ([12]: Sects. 65–75); a detailed *Übersicht* of their variegated uses within specific language games will make that manifest.

5.3 Firth's "Context of Situation" and "Collocation"

What links Word2vec with Wittgenstein's philosophical insights about the meaning of a term as its rule-governed use within a particular language game is the application which these philosophical ideas received in the linguistic theory of English linguist John Rupert Firth.

The breakpoint for contemporary general linguistics was the publication of Ferdinand de Saussure's *Course in General Linguistics* in 1916. One of de Saussure's most important ideas was to consider language as a system of signs (as compared to systems in many other fields). Linguistic sign is constituted of the signifier, or sound pattern, and the signified, or mental concept. Language signs belong to the language as a system, so that a change in any sign affects the system as a whole [9]. A couple of decades later, Firth expanded on de Saussure's conception of the linguistic sign: signs are not only dependent on the language system, but also their meaning can change with the context in which they are used.

Firth has repeatedly stressed that linguistics should not abstain from addressing the question of meaning ([3]: 190) and that "the complete meaning of a word is always

[3]In their analytical commentary on *Philosophical Investigations*, Baker and Hacker suggest that this technical term of Wittgenstein's should be translated with the English word "surview" ([1]: 531–545).

contextual, and no study of meaning apart from a complete context can be taken seriously" ([3]: 7). Contextual considerations must include "the human participant or participants, what they say, and what is going on" ([3]: 27), since "language is a way of dealing with people and things, a way of behaving and making others behave" ([3]: 31); language is used by persons in a social environment ([3]: 187), insists Firth, with the aim of maintaining a certain "pattern of life" ([3]: 225). The fundamental notion in Firth's linguistic theory is that of the *context of situation*, which he acknowledges inheriting from his collaborator, anthropologist Bronisław Malinowski ([3]: 181). A context of situation is specified when the following is known: (1) the verbal and the non-verbal actions of the participants, (2) the objects involved, and (3) the effects of verbal action ([3]: 182). Firth immediately stresses the similarity between his concept of context of situation and Wittgenstein's notion of the language game. He approvingly cites Wittgenstein's dictum that "the meaning of words lies in their use" ([2]: 179) and that "a language is a set of games with rules or customs" ([2]: 139). The notion of the context of situation is meant to emphasise the social dimension of language. Just as Wittgenstein, and Frege before him, Firth argues against mentalistic accounts of meaning: the theory of Ogden and Richards [8], influential in his time, which identifies meaning with "relations in a hidden mental process" ([3]: 19), he considers as an unacceptable remnant of Cartesianism. Again, entirely in tune with the later Wittgenstein is his forsaking the universal theory of language in favour of a descriptive and detailed study of what he calls "restricted languages", i.e. languages scaled down to particular contexts of situations; the examples of restricted languages he provides are: air-war Japanese, Swinburnese lyrics or modern Arabic headlines ([2]: 29).

In further analysing the contextual meaning of a word, Firth distinguishes its many dimensions and singles out the one most eligible for empirical investigation: collocation. Collocation is "quite simply the mere word accompaniment, the other word-material is which [the word is] most commonly or most characteristically embedded" ([2]: 180). The idea is that if two words have different accompaniments, they are already semantically distinguishable by that feature alone. To cite Firth's example: it is evident that "cow" does not mean the same as "tigress" since "cow" appears in collocations such as "They are milking the cows", while "tigress" does not ([2]: 180). In introducing the notion of collocation, Firth paraphrases Wittgenstein in asserting that "a word in company may be said to have a physiognomy" ([3]: xii). A non-Wittgensteinian move, however, which Firth made, and which significantly paved the way for Word2vec, is that he explicitly declared collocation—a limited excerpt of the purely linguistic element of the context—to be a part of the word's meaning [3]: 196); his famous and most quoted line is: "You shall know a word by the company it keeps" ([2]: 179). It is evident that the vector which Word2vec associates with a word as its meaning is designed to capture its collocations.

5.4 Word2vec

It is straightforwardly understandable that the concepts *beer* and *wine* are more similar to each other than the concepts *beer* and *cat*. A possible explanation for this being so is that the words for these concepts, "beer" and "wine", appear in the same contexts more often than the words "beer" and "cat" do. This way of thinking is in the background of Word2vec model of word embedding, and it is called the distributional hypothesis.[4] Here, neural networks are used to recognise such similarities.

Neural networks, or more precisely, artificial neural networks, such as those used in Word2vec models for word embedding,[5] are computing systems intended to emulate the functionality of the (human) brain. A model most similar to the human brain would be a computer system that processes numerous data in parallel. Both generally accepted models of computing—von Neumann and Harvard architectures— greatly differ from the concept of neural network: from building block types to the number of "processors", connections and information type.

In the early beginnings of AI research, two models emerged—the symbolic and the connectionist. The symbolic approach tends to aggregate specific domain knowledge with a set of atomic semantic objects (symbols) and to manipulate those symbols through algorithms. Such algorithms, in real-life applications, have almost always an NP-hard complexity or worse, rendering massive search sets in problem-solving. This makes the symbolic approach suitable for certain restricted artificial use cases only. On the other hand, the connectionist approach is based on building a system with internal architecture like that of a brain, which "learns" from experience rather than have a preset algorithm to follow. It is used in numerous practical cases, which are too difficult for the symbolic approach; it is applied in the domain of formal languages for solving: the string-to-string correction problem, the closest string problem, the shortest common supersequence problem, the longest common subsequence problem, etc.

Several critical differences between the paradigms of classic computation architecture and (artificial) neural network can be displayed in the Table 5.1.
Thus, a neural network can be roughly defined as a set of simple interconnected processing elements (units, nodes), whose functionality is based on the biological neuron used in the distributive parallel data processing. It is purposely designed for answering the problems of classification and prediction, that is, all problems that have a complex nonlinear connection between input and output. It is significantly advanced in solving assessments of nonlinearity, robust on data errors, highly adap-

[4]One of the first formulations of the distributional hypothesis is often associated with the already mentioned Firth's ([2]: 179) dictum "You shall know a word by the company it keeps". More precise, and anticipating Word2vec, was Harris's claim: "All elements in a language can be grouped into classes whose relative occurrence can be stated exactly. However, for the occurrence of a particular member of one class relative to a particular member of another class it would be necessary to speak in terms of probability, based on the frequency of that occurrence in a sample." ([6]: 146).

[5]Word embedding is a process in which semantic structures (words, phrases or similar entities) from a certain vocabulary are mapped to and mathematically modelled as Euclidean vectors of real numbers.

Table 5.1 Differences between the paradigms of classic computation architecture and (artificial) neural network

Standard computing architecture	(Artificial) neural network
Predefined detailed algorithms	Self-sustained or assisted learning
Only precise data is adequately processed	Data can be unclear or fuzzy
Functionality dependable on every element	Processing and result are not largely dependable on a single element

tive and capable of learning; it works well with fuzzy or lossy data (from various sensors or non-deterministic data) and can work with a large number of variables and parameters. As such, it is beneficiary for pattern sampling, processing of images and speech, optimization problems, nonlinear control, processing of imprecise and missing data, simulations, the prognosis of time series and similar uses. Artificial neural networks generally work in two phases: learning (training) and data processing.

Word2vec suitability for semantic similarity is based on the implementation of word representations, which rests on the aforementioned distributional hypothesis. In other words, the context, in which a word is used, is provided by its nearby words. The goal of such representations is to capture the syntactic and semantic relationship between words.

As model examples, Word2vec uses two similar neural network-based distributional semantic models for generating word embeddings—CBOW (Continuous Bag-of-Words) and Skip-gram. Tomas Mikolov's team from Google created both models in 2013. CBOW attempts to predict the current word based on the small context window surrounding that word. CBOW suggests a concept where the projection layer is shared between all words and the nonlinear hidden layer is removed; the word distribution and order in the context do not influence the projection. This model also proved to be of substantially lower computational complexity. Skip-gram architecture is similar, but instead of predicting the current word on the basis of its context, it tries to predict the word's context with respect to the word itself. Therefore, the Skip-gram model intends to find word patterns that are useful for predicting the surrounding words within a specific range in a sentence. Skip-gram model estimates the syntactic properties of words slightly worse than the CBOW model. Training of the Skip-gram model does not involve dense matrix multiplications, which makes it extremely efficient [7]. Let us consider a simple example. For the words "cat", "wine" and "beer", we have the following vectors:
vec ("cat") = (0.1, 0.5, 0.9)
vec ("wine") = (0.6, 0.3, 0.4)
vec ("beer") = (0.5, 0.3, 0.3)
As can be seen, the vectors corresponding to the words "wine" and "beer" have more similar values than any of them has to the vector for the word "cat", hence the concepts expressed are more similar to each other than to the concept cat. We

can presuppose that the second value in the vectors for "wine" and "beer" stands for a feature like *is an alcoholic drink*. The similarity in meaning between words can be calculated as the cosine of the angle between vectors. In order to train another model using the already mentioned representations, we can feed them into a different machine learning model. Values assigned to each word are the result of the Skip-gram model, which has a role in determining which words often appear in similar contexts. In case two words can often be found surrounded by similar other words, their resulting vectors will be similar and the cosine of their angle will approach 1. Therefore, the word vector can be regarded as a compressed representation of its contexts of use. When the whole process is over, each word in the dictionary has been assigned its vector representation, and those representations can be listed alphabetically (or otherwise): the results will be N-dimensional vectors, where N is the number of words in the whole vocabulary. Additional technical details would exceed the scope of this chapter.

5.5 Conclusion: Differences Between Wittgenstein's Understanding of Word Meaning and that Facilitated by Word2vec

One of the chief differences between Wittgenstein's and the Word2vec approach to word meaning lies in the fact that although both emphasise the importance of context, its scope is differently understood. Word2vec offers a restricted view of what constitutes a context of use of a certain word: it is limited to directly neighbouring words only; neighbouring phrases with several words are not considered, let alone whole sentences in which they occur. Such a determination of context, even if derived from Firth's notion of collocation, seems too austere in comparison to its linguistic ancestor: the examples of collocation, as given by Firth, are typically more complex and often include sentences. The divergence from Wittgenstein is even starker: a meaningful sentence was his narrowest understanding of context, which was in the transitional period replaced with the wider *Satzsystem*, and, still later, with an even more encompassing language game. However, this simplification of context and restriction to a single mode of meaning—collocation—enabled the creation of such a widely applicable formalisation as Word2vec.

In the *Tractatus* period, Wittgenstein believed that all sub-sentential expressions, at least when the sentence is fully analysed, have the role of names ([11]: Sect. 4.22), and that their meaning is fully exhausted in their reference (*Bedeutung*)—an extra-linguistic object (*Gegenstand*) that they stand for in the context of the sentence. As names, in his view, refer directly, without the mediation of sense (*Sinn*) ([11]: Sect. 3.142), they have no linguistic meaning at all. According to Wittgenstein in the *Tractatus*, meaning is a relation between names and their bearers; it does not connect words with other words of the same language ([11]: Sects. 3.202, 3.22).[6] Although Wittgenstein later changed his mind and criticised his early views on this matter ([12]: Sect. 38), he never denied that at least *some* expressions in language have a referential role: it is part of their meaning to aim to pick out something in the extra-linguistic reality and represent it within a language game. The referential aspect of meaning, however, eludes Word2vec. Vectors delivered by Word2vec track only collocations, they register exclusively inter-linguistic connections between words.

Throughout his philosophical development, Wittgenstein thought that words have meaning only in a wider context in which they are used, but he did not identify the context with the meaning. A word has meaning inside a language game, but from that it does not follow that its meaning is the language game itself, or an exhaustive list of all the various language games the word is employed in. The Wittgensteinian conception of word meaning would be better represented with a function that to every language game (in which the word is used) ascribes the meaning the word has in that particular game: a set of rules governing the use of the word in the game. If the meaning of a word consisted simply in the list of contexts it is used in, then any two words employed in the same language games would automatically turn out synonymous, even if their uses were governed by different semantic rules in each game. Specifying the contexts in which a word is applied tells us something important about the word's meaning, but it falls short of its full account: it still does not amount to Wittgenstein's *Übersicht*. This is especially so since words that belong to the same semantic category have a tendency to occur in the same linguistic surroundings, even though their meanings may vary considerably.

In Word2vec, every word is assigned a *unique* vector which codifies all its collocations and thus represents its meaning. Consequently, if two words are such that there is a context in which one of them cannot be substituted with the other, their Word2vec vectors will, expectedly, be different. However, it may happen that some words do not have the same overall meaning but are synonymous within some specific contexts: one of them may not be used in all the language games in which the other is, but still in some language games they may be applied according to the same rules. To borrow an example from Firth ([2]: 179), in utterances like "You silly ass!" and "Don't be such an ass!" the word "ass" is used synonymously with the word "fool", although these words have thoroughly different meanings in some other contexts and cannot be interchanged therein, for instance, in the utterance "An ass has longer ears than a horse." Such cases of synonymy-relative-to-a-context cannot be accounted for in Word2vec, precisely because Word2vec does not operate with the notion of meaning in a particular context, but instead identifies the meaning of a word with a list of contexts (understood as collocations). With his concept of *Übersicht*, as a perspicuous presentation of (potentially divergent) meanings a word has in separate contexts, Wittgenstein, on the other hand, has a powerful enough tool to elucidate cases of local synonymy.

---

[6]If the meaning of a word is explicated by means of a definition, such word cannot be a name, since names are semantically simple, but only an abbreviation for the definition in question.

Wittgenstein's extremely influential distinction between words which have sharp and clear necessary and sufficient conditions of application, on the one hand, and words for family resemblance concepts, on the other, is also inexpressible in Word2vec. An *Übersicht* of different uses a word is put to in various language games is supposed to transparently exhibit if there is a common thread which runs through all these uses or not; this, however, cannot be read off the Word2vec vectors. Vectors that correspond to words denoting family resemblance concepts, such as "game", are not in any apparent way distinct from vectors that correspond to sharply defined words: in both cases, the vectors only register the contexts of use and are silent on whether there is a shared core meaning in all these contexts. The inability to articulate this Wittgensteinian distinction in Word2vec is another consequence of the fact that Word2vec does not map contexts (in which a word is used) to meanings (the word has in each of them), but conflates them.

To conclude, although Word2vec and similar models are a significant step forward in natural language understanding, as it is clear from the discussion above, there are still a lot of components contained in the meaning of words in natural language that cannot be captured using this method. There is no doubt that, in recent years, neural networks have made an important improvement in natural language processing, including their use in machine translation. On the other hand, artificial intelligence still does not have the capacity to go deep inside the problems of meaning in general, especially as far as specific features of natural language, such as synonymy, are concerned.

Acknowledgements Andrej Jandrić's research has been supported by the Ministry of Education, Science and Technological Development of the Republic of Serbia (179067) and the University of Rijeka, Croatia (uniri-human-18–239).

References

1. Baker GP, Hacker PMS (1980) Wittgenstein: understanding and meaning. Basil Blackwell, Oxford
2. Firth JR (1968) Selected papers 1952–1959. Longmans, Green and Co Ltd., London and Harlow
3. Firth JR (1969) Papers in linguistics 1934–1951. Oxford University Press, London
4. Frege G (1884/1953) The foundations of arithmetic. A logico-mathematical enquiry into the concept of number. Basil Blackwell, Oxford
5. Frege G (1892/1960) Concept and object. In: Translations from the philosophical writings of Gottlob Frege. Oxford, Basil Blackwell, pp 42–55
6. Harris ZS (1954) Distributional structure. Word 10(2–3):146–162. https://doi.org/10.1080/00437956.1954.11659520
7. Mikolov TS, Sutskever I, Chen K, Corrado G, Dean J (2013) Distributed representations of words and phrases and their compositionality. In: Advances in neural information processing systems, neural information processing systems foundation, Inc., pp 3111–3119
8. Ogden CK, Richards IA (1923) The Meaning of meaning: a study of the influence of language upon thought and of the science of symbolism. Harcourt, Brace and World Inc., New York
9. de Saussure F (1916/1959) Course in general linguistics. Philosophical Library, New York
10. Shanker SG (1987) Wittgenstein and the turning-point in the philosophy of mathematics. State University of New York Press, New York

11. Wittgenstein L (1922/1999) Tractatus logico-philosophicus. Dover Publications Inc., Mineola, New York
12. Wittgenstein L (1953/2001) Philosophical investigations. Blackwell Publishing, Oxford
13. Wittgenstein L (1958/1997) The blue and brown books: preliminary studies for the 'Philosophical investigations'. Blackwell, Oxford
14. Wittgenstein L (1964/1998) Philosophical remarks. Basil Blackwell, Oxford
15. Wittgenstein L, Waismann F (2003/2013) The voices of Wittgenstein: the Vienna Circle. Routledge, London and New York

Chapter 6
Rudolf Carnap–The Grandfather of Artificial Neural Networks: The Influence of Carnap's Philosophy on Walter Pitts

Marko Kardum

Abstract The importance and relevance of philosophy for the development of the AI is often neglected. By revealing the influence of Rudolf Carnap on Warren McCulloch's and especially Walter Pitts' work on artificial neural networks, this influence could be reexposed to the scientific community. It is possible to establish a firm connection between Rudolf Carnap and Walter Pitts by pointing out to a personal relationship but also to a more internal structure of that influence as evidenced by Pitts' usage of Carnap's logical formalism. By referring to Carnap's work, Pitts was able to abide Kantian notion of unknowable and undescribable and to lay foundation of the world as a completely describable structure. It also meant that it could be possible to construct machines that use neural networks just as the biological entities do. Thus, Carnap could be regarded as the grandfather of artificial neural networks and logic, divided by the unfortunate historical development, could become united again as a single discipline that keeps both it's mathematical and philosophical side.

Keywords Carnap · Pitts · Artificial neural networks · Connectionism · Neurocomputational formalism

Artificial intelligence (AI) is a propulsive modern and contemporary field connected to many different research and scientific areas, including both science and humanities. As such, it presents a field of major interest to philosophers, mathematicians, engineers, computer programmers, etc. In common and maybe even most accepted views today, connection between AI and different scientific fields is most often regarded as a completely plausible and almost a natural one. Yet, its connection to philosophy and especially the fact that one can justifiably claim that AI originates from philosophy is more often than not perceived as a vague effort undertaken by those who want to appropriate this propulsive field for themselves while having no true merit

The present research was supported by the short-term Grant *Philosophical Aspects of Logic, Language and Cybernetics* funded by the University of Zagreb under the Short-term Research Support Program.

M. Kardum (✉)
Faculty of Croatian Studies, University of Zagreb, Zagreb, Croatia
e-mail: mkardum@hrstud.hr

S. Skansi (ed.), *Guide to Deep Learning Basics*,
https://doi.org/10.1007/978-3-030-37591-1_6

for it. Thus, the main goal of this chapter will be to show that philosophy has every right to claim its relevance for the development of the AI (the question of importance of philosophy for the development of the AI not only as a part of its history but as a relevant discipline in its current development will be put aside for now) and this task will be done by showing the influence of one of the famous philosophers, that is Rudolf Carnap's influence, on Walter Pitts' development of artificial neural networks.

The importance of philosophy could be stressed even more as some of the "core formalisms and techniques used in AI come out of, and are indeed still much used and refined in, philosophy" [6], although extensive tracking of philosophical roots in AI development can go all the way back to Descartes and even Aristotle [25]. AI is usually described as a field which is concerned with constructing artificial creatures that act in an intelligent way and may those creatures be regarded as artificial animals or even as artificial persons, it represents, as such, a field of a major interest to philosophers. Answering the question whether these creatures, in different contexts, are artificial animals or even artificial persons may be well worth on its own but do not represent an important insight into our topic. However, a deeper insight into Carnap's influence on Walter Pitts' work might reveal a continuous relation between philosophy and AI and this relation might prove to be a missing link which connects science and humanities within the development of AI.

When considering the history and early development of AI, it is impossible not to mention famous conference of summer 1956 at Dartmouth College, in Hanover, New Hampshire. The conference was sponsored by DARPA (Defense Advanced Research Project Agency) and was attended by "John McCarthy (who was working at Dartmouth in 1956), Claude Shannon, Marvin Minsky, Arthur Samuel, Trenchard Moore (apparently the lone notetaker at the original conference), Ray Solomonoff, Oliver Selfridge, Allen Newell, and Herbert Simon" [6]. Among other notable conclusions, the conference remains well known as the birthday and the birthplace of the term AI. However, it would be hard to defend the claim that nothing of the field of AI, besides the of the field name itself, did not exist before 1956 and the aforementioned conference. Maybe the two best-known examples of development of the rich AI field before the term was even coined are works of Alan Turing and Walter Pitts (together with Warren McCulloch) which greatly improved our understanding of machine learning and problem-solving. In other words, their work led to the development of the AI field which was trying to build a machine that could actually think and learn. Turing [29] was interested in giving a systematical analysis of algorithms which function as mechanical instructions for each phase of machine problem-solving.[1] Known as the development of the Turing machine, this famous analysis was just one step along the way of understanding machine learning. The next step was to consider the possibility of existence of machines that could actually think. The discussion whether machines (or later, computers) can really replicate or just mimic human thought was, in a way, anticipated by Turing. In his famous paper published in Mind [30], he suggested

[1]For informal description of Turing machines see Rescorla [24] and for its rigorous mathematical model see De Mol [11].

the imitation game , a mind experiment which is known today as the Turing test (TT),[2] which will, in a way, force us to replace the question "Can a machine think?" with a more precise one that can actually produce some meaningful results—"Can a machine be linguistically indistinguishable from a human?" [6]. By switching our focus from the definitions of words *"machine"* and *"think"*, we avoid the statistical nature of the possible answer to the stated question "Can a machine think?" due to different common language usage of those terms [30].

The TT can also be interpreted as Turing's suggestion and attempt to overthrow a demand for building machines that have the full mental capacities of humans and to replace it with machines that only appear to have them [26]. In this way, Turing again anticipated the dispute over "strong" and "weak" AI arguments. Computer science today, although differing somewhat from the Turing's simplified model, is based on Turing's work and is making a rapid progress in developing more complex computing systems. However, what is very interesting for us is that Turing in in TT also suggested how to construct such machines:

> He suggests that "child machines" be built, and that these machines could then gradually grow up on their own to learn to communicate in natural language at the level of adult humans. This suggestion has arguably been followed by Rodney Brooks and the philosopher Daniel Dennett (1994) in the Cog Project. In addition, the Spielberg and Kubrick movie A.I. is at least in part a cinematic exploration of Turing's suggestion. [6]

Machines gradually growing up and learning how to communicate is what brings us to the work of Walter Pitts, especially his work on artificial neural networks. This brings us to the end of short review of AI, machine learning and neurocomputing and their relation.

Throughout history, artificial neural networks were known by various names. Some of those names are cybernetics, nerve nets, perceptrons, connectionism, parallel distributed processing, optimization networks, deep learning and, of course, artificial neural networks (ANN). The latter will be used in this chapter. ANN is connectionist[3] and as such part of computing[4] systems inspired by and interpreted as biological neural networks. It is important to emphasize that ANN are not identical to biological neural networks that constitute animal (including human) brains, but there is a strong resemblance between them. Although the goal of using ANN was to develop neural network system that could approach and solve general and complex problems in a way similar to the way a human brain does, it was dealing with specific tasks that proved to be a more realistic approach and offered better results which could be applied in different areas such as video games, medical diagnosis, speech recognition, and even painting.

---

[2]For objections to the TT see Block [4] and for more detailed overview of the TT see Oppy and Dowe [17].

[3]Connectionism is part of cognitive science that explains intellectual abilities and learning using artificial neural networks. For further reading on connectionism see Buckner and Garson [7].

[4]Computational theory of mind holds that the mind itself is a computational system or a thinking machine. For further reading on computational theory of mind see Rescorla [24].

From the very onset and first attempts toward the grand original goal, all neural network models diverged from biology and the biological brain. Nevertheless, ANN retained strong resemblance to biological brain in terms of connecting units that work as biological neurons and that can, as the synapses in a biological brain, transmit and receive signals and then process and "inform" other neurons connected to it by signaling them. This new form of non-logicist formalism treated the brain as a computational system or, more precisely, it tried to explain intelligence as non-symbolic processing like the one that can be found at some level (at least at the cellular level) in the brain. Having that in mind, it is safe to say that it was a new paradigm in understanding and creating AI that triggered the race between the symbolicist and connectionist approach to AI. Connectionist paradigm developed the following neurophysiology rather than logic and has greatly influenced later development of computer science. The pioneer work, as it is often emphasized, was done by Warren S. McCulloch and Walter Pitts [16] in their famous paper[5] *A Logical Calculus of the Ideas Immanent in Nervous Activity* where they "first suggested that something resembling the Turing machine might provide a good model for the mind" [24], although their proposal of artificial neural networks differed significantly from the Turing machines.

However, there are views that, despite its significance, McCulloch's and Pitts' paper work remained relatively underrated both historically and philosophically. That seems strange enough by itself and becomes even more strange considering there were already biophysicists that were engaged in mathematical investigations of (biological) neural networks at that time as [19] notes.

Furthermore, there was no similar theory developed at the time and instead of producing seminal results in neurobiology, McCulloch's and Pitts' work influenced far more the field that will later be known as AI [15]. So, what was the contribution of McCulloch's and Pitts' work regarding what was discussed earlier and the process of machines' learning? Let us here quote the opening passage from their famous paper:

> Because of the "all-or-none" character of nervous activity, neural events and the relations among them can be treated by means of propositional logic. It is found that the behavior of every net can be described in these terms, with the addition of more complicated logical means for nets containing circles; and that for any logical expression satisfying certain conditions, one can find a net behaving in the fashion it describes. It is shown that many particular choices among possible neuropsychological assumptions are equivalent, in the sense that for every net behaving under one assumption, there exists another net which behaves under the other and gives the same results, although perhaps not in the same time. Various applications of the calculus are discussed. [16, p. 99]

Given the quoted passage, it can be asserted along the lines of [19] that their major contribution should be divided into four parts: (1) constructing a formalistic approach which eventually led to the notion of "finite automata",[6] (2) an early technique of

[5]Just as illustration of the influence of their paper, the interested reader can consult the excellent handbook edited by Michael A. Arbib The Handbook of Brain Theory and Neural Networks [2]—it is almost impossible to find any article that is not referring to McCulloch's and Pitts' paper.

[6]A finite automaton is an abstract machine constructed through a mathematical model of computation that can be in exactly one finite state at a time and this state can change depending on external

logic design, (3) the first use of computation for resolving some mind–body problems, and (4) the first modern theory of mind and brain.

However, here we can summarize their contribution as a development of a non-logistic, connectionist, and neurocomputational formalism that enhanced machine learning and developed a brain-like organization of AI as much as the *brain-as-a-computer* paradigm. According to McCulloch and Pitts [16], it should be added to the fact that all our theories or ideas, as well as sensations, are realized by the activity of our brains, and that the same network determines epistemic relations of our theories to our observations and from these to the facts of the world. Thus mental illness such as delusions, hallucinations, and confusions represent alterations to the network and empiric confirmation that if our networks are undetermined, our facts are also undetermined and there is no observation, sensation, or theory that we can get hold of. The final consequence of this, as they put it, is a somewhat Kantian sentence and the final dismissal of the metaphysical residue of our knowledge:

> With determination of the net, the unknowable object of knowledge, the "thing in itself," ceases to be unknowable. [16, p. 113]

When he was asked about neural modeling (cf. [1]), Jack D. Cowan saw computer technology as being a driving force in applying theory to real-world problems. He saw development of artificial neural networks as crucial to this application, even when there were obvious problems of neural network approaches to language. His view on McCulloch's and Pitts' work is of great significance:

> It's very like the content of the McCulloch-Pitts paper itself. The late Donald Mackay, whom I knew very well, characterized their theorem as follows: if you are arguing with someone about what a machine can or cannot do, and you can specify exactly what it is the machine cannot do, then their theorem guarantees that there exists at least one machine that can do exactly what you said it cannot do. If you can specify in enough detail what it is that you say a machine can't do, you've produced a solution. So the real question is, "Is everything out there describable?" [1, p. 125]

It is now distinctly possible to say that McCulloch and Pitts considered neural networks to be able to teach machines to perform every action describable or, to use their own words, every action that is defined, that is, every action that is determined by defining our net the brain. To define it, we need to describe what it does and, even when dealing with a psychic unit *psychon*, we can reduce its actions to activity of a single neuron which is inherently propositional. Hence, all psychic events have intentional or "semiotic character" and "in psychology, introspective, behavioristic or physiological, the fundamental relations are those of two-valued logic" [16, p. 114]. So, how is all of this related to Rudolf Carnap's work?

In their famous paper, McCulloch and Pitts referred only to three works of other authors and those are Hilbert's and Ackermann's *Grundüge der Theoretischen Logik*

---

inputs or conditions that have been met. It is similar to the Turing machine, although it has less computational power because of the limited number of finite states and, consequently, limited memory. Some of the most known and simple finite automata examples are vending machines, elevators, and traffic lights.

[14], Whitehead's and Russell's *Principia Mathematica* [32] and Carnap's *The Logical Syntax of Language* [9]. And while Hilbert's and Ackermann's and Russell's and Whitehead's work was already famous, and thus might have been expected to be referred to by McCulloch and Pitts, Carnap's work was not quite there yet and inclusion of his book came as a bit of surprise. That fact alone says enough about Carnap's influence on development of ANN (as the authors perceived it) but let us go one step further in explaining this influence.

Let us start by shortly sketching the characters of McCulloch and Pitts.[7] As pointed out in several testimonials by their friends and colleagues in Anderson's and Rosenfeld's [1] noteworthy book which, by interviewing some of the scientists involved, describes more closely the beginnings of development of AI and ANN, the atmosphere and relations among most prominent names of the field, Warren McCulloch is often characterized as a generous and outgoing person who had taken care of his younger colleagues and had even housed Pitts and Lettvin during their early days. He was mostly recognized as a creative and imaginative force among the members of his neurophysiology group. His creativity and imagination went so far that he was considered by Jack D. Cowan as the most eccentric one among Pitts, Lettvin, and Wall. Turing, who at least once met McCulloch, even thought of him as a charlatan. The imaginative driving force of McCulloch had its opposite in the character of Walter Pitts. Pitts was in a way, as Lettvin, McCulloch's protégé but was also closely connected to Norbert Wiener for whom he started to work in 1943 and of whom he actually thought of as a father figure which he never really had. That is the reason why he felt being left in the middle between McCulloch and Wiener, had a nervous breakdown from which he never recovered and started to destroy his own work. Especially traumatic was the nervous breakdown Pitts had after McCulloch's and Wiener's dispute. This was devastating since Pitts was, as witnessed by Cowan, considered to be the brain of McCulloch's group:

> I was very much impressed with Pitts and his insights. Walter was really the intelligence behind Lettvin and McCulloch. I think it was Walter who was the real driving intelligence there. Since 1921 Warren had had an idea of somehow writing down the logic of transitive verbs in a way that would connect with what might be going on in the nervous system, but he couldn't do it by himself. In 1942, he was introduced to Pitts, who was then about seventeen years old. Within a few months Walter had solved the problem of how to do it, using the Russell-Whitehead formalism of logic, which is not a transparently clear formalism. Nonetheless, they had actually solved an important problem and introduced a new notion, the notion of a finite-state automaton. So here was this eccentric but imaginative Warren and this clever young Walter doing this stuff together. [1, p. 104]

What is really interesting and is often told about Pitts is his path that led him to McCulloch. Lettvin, who was considered Pitts' best and inseparable friend, is the most reliable source to document Pitts' life. According to him Lettvin (cf. [1, pp. 2–12]), Pitts was an autodidact who taught himself mathematics, logic and a fair number of languages and ran away from home when he was 15. Maybe the most frequently mentioned episode of his early life includes Whitehead's and Russell's *Principia*

[7] The best references about their life and work are Anderson and Rosenfeld [1] for both of them, Arbib [3] for McCulloch and Smalheiser [27] for Pitts.

Mathematica. After being chased by bullies, he hid himself in a public library where he found it and read it in a three-day period after which he sent a letter to Russell, pointing out some problems in the book he considered to be serious. Russell invited him to go and study in England but Pitts refused and went to the University of Chicago where he attended some lectures but never registered as a student. There, in 1938, he read Carnap's new book The Logical Syntax of Language and did almost the same thing as with Principia Mathematica. Without even introducing himself, he walked into Carnap's office and once again pointed at some problems and flaws in the book and left without saying a word about himself.[8]

Of course, Carnap have had tough time finding him, but succeeded in the end and managed to persuade University of Chicago to give Pitts a menial job.[9] At the time, Pitts was considered to be Carnap's protégé.[10]

This is where we find proof about Carnap's influence on Pitts. It is for sure beyond any doubt that Pitts knew about Carnap's work very well since his early days at the University of Chicago. Furthermore, as confirmed by his colleagues, he was later attending Carnap's and Russell's lectures in logic and there he got to know their work to details. In combination with Pitts' intelligence and influences as evidenced, it is safe to say that McCulloch's and Pitts' work[11] on ANN is influenced by Carnap. And that influence is, surely, most notable in their usage of

> "the uncommon logical formalism of Carnap (1938) and Hilbert and Ackermann (1927) for the presentation of their results" [18, pp. 230]

which was, undoubtedly, acquired by Pitts. So, the next logical step would be to discover and describe properties of Carnap's logical formalism that can be found in McCulloch's and Pitts' work on ANN. However, let us first see why logic played such important role for Pitts in modeling ANN.

While studying with Carnap, Pitts wrote three papers on neuron network modeling[12] that preceded the famous paper written with McCulloch. What is usually considered to have been a trigger for this paper was discovery of inhibitory synapses

[8]Rudolf Carnap confirmed that this event actually happened and had explicitly said that after Pitts made his objections some parts of his own book were unclear even to him [5].

[9]Another interesting insight into Pitts' life and career development from homeless young boy to one of the masterminds of early AI development is Smalheiser's paper on Walter Pitts [27].

[10]As a note on Pitts' life, it is interesting to read how McCulloch later on was informing Carnap about Pitts' progress and achievements: *"He is the most omniverous of scientists and scholars. He has become an excellent dye chemist, a good mammalogist, he knows the sedges, mushrooms and the birds of New England. He knows neuroanatomy and neurophysiology from their original sources in Greek, Latin, Italian, Spanish, Portuguese, and German for he learns any language he needs as soon as he needs it. Things like electrical circuit theory and the practical soldering in of power, lighting, and radio circuits he does himself. In my long life, I have never seen a man so erudite or so really practical."* [13, p. 60].

[11]It has to be said that it applies not only to "The Logical Calculus of the Ideas Immanent in Nervous Activity" but also to their second paper "How we know universals" [23].

[12]These papers are "Some observations on the simple neuron circuit" [21], "The linear theory of neuron networks: The static problem" [20] and "The linear theory of neuron networks: The dynamic problem" [22].

that, together with Pitts' knowledge of propositional logic and McCulloch's knowledge of neurophysiology, led to development of the McCulloch–Pitts neuron model [27, p. 219]. Although, as witnessed by Pitts' best friend Lettvin [1, p. 3], McCulloch and Pitts got together on "The Logical Calculus of the Ideas Immanent in Nervous Activity" in the evening on the same day they moved in with McCulloch and his family, it would not be strange to say that Pitts, back then an 18-year-old boy, had a major influence on developing new approach to neuron modeling. The most obvious reason to think so is, as already expressed in Donald Mackay's words about McCulloch's and Pitts' theorem, Leibniz influence on Pitts and his view that a logical machine could do anything that can be completely described:

> Walter at that time was, if I remember correctly, about eighteen, something like that. Walter had read Leibniz, who had shown that and task which can be described completely and unambiguously in a finite number of words can be done by a logical machine. Leibniz had developed the concept of computer almost three centuries back and had even developed a concept of how to program them. I didn't realize that at the time. All I knew was that Walter had dredged this idea out of Leibniz, and then he and Warren sat down and asked whether or not you could consider the nervous system such a device. So they hammered out the essay at the end of 42. [1, p. 3]

The notion of completely describable tasks, taken from Leibniz, was supported by McCulloch–Pitts neurons model which "receive[s] a set of monosynaptic excitatory and inhibitory inputs and that fire whenever the net sum exceeds a threshold" [27, p. 219] or otherwise do not fire. This led Pitts to believe that, by mastering logic, neural nets became completely determined.

In turn, this could get hold of our innate structure and transfer it to the artificial level. Thus, as already mentioned as a Kantian sentence in McCulloch's and Pitts' paper, the world stops being indescribable and therefore unknowable. This view is supported by Cowan's understanding of McCulloch's group:

> All through the McCulloch group was this idea that there was an innate structure there. They believed in the Kantian notions of synthetic a priori. That's the kind of thinking that led Lettvin and Pitts to come up with "What the Frog's Eye Tells the Frog's Brain". [1, p. 108]

However, this is where Pitts' certainty about logic as being the key to the world ends:

> See, up to that time, Walter had the belief that if you could master logic, and really master it, the world in fact would become more and more transparent. In some sense or another logic was literally the key to understanding the world. It was apparent to him after we had done the frog's eye that even if logic played a part, it didn't play the important or central part that one would have expected. And so, while he accepted the work enthusiastically at the same time it disappointed him. [1, p. 10]

So, where does Rudolf Carnap fit in? Besides being Pitts' professor at the University of Chicago, besides being a target of Pitts' criticism and a man who helped Pitts stay at the University of Chicago, it has already been pointed out that Carnap's book was one of the three books referred to by McCulloch and Pitts in their famous paper. It has also been pointed out and shown that Carnap most surely influenced Pitts more than McCulloch. Pitts attended Carnap's lectures and was determined to show that by mastering logic one would be able to completely understand the world. In fact, this understanding had to come from Carnap's work considering that "Pitts immediately saw how one might apply Carnap's formalism to McCulloch's ideas about the logic of the nervous system" [3, p. 197] and that "the use of logical formalism by McCulloch and Pitts is a clear consequence of their conviction that in physiology the fundamental relations are those of two-valued binary logics" [18, p. 230].

As McCulloch's group did, Carnap also starts with a Kantian question: how is mathematics, both pure and applied, really possible [12]? He avoids both pure reason and naive empiricism and establishes logical empiricism in which mathematics and logics are not a part of empiricism nor a part of pure intuition. For Carnap [9], it is crucial to form *analytic* a priori sentences that are true only by virtue of their constituent terms and that require no empirical evidence whatsoever. On the other hand, scientific sentences are *synthetic* a posteriori and prove to be true or false depending on the state of objects in the real world.

For Carnap, this offers a satisfactory methodological analysis of science and bypasses Gödel's incompleteness theorems, leaving its results intact and not trying to prove the contrary. These *synthetic* a posteriori sentences, dependent on empirical evidence, can be expressed in any natural language and, by the principle of tolerance, *analytic* a priori sentences of mathematics and logic can be formulated in any way that can prove them to be true or false by aforementioned virtue of their constituent terms. This way, Carnap's (conventional) formalism[13] leaves the possibility of reducing all of the complexity of arithmetic to logic in a Russell–Whitehead manner and not being shaken by Gödel's results. By managing that, Carnap secures a satisfactory methodological analysis of science and holds that logic, as metalanguage, and natural language that we use to communicate our empirically gathered results form a logical syntactic order in which propositions secure the truth of scientific sentences about our real world.

This model rejects every kind of dogmatic claims, namely, those of metaphysics, and provide a safe progress of our knowledge. It is here where we find the true influence of Rudolf Carnap on Walter Pitts[14] if we get firm hold of the truth of our propositions analytically and a priori and "manage to map them in the brain" [3, p. 4], as well as secure the truth of our a posteriori sentences by submitting them to rigorous analysis of our propositions, no matter how we construct them to work as long as they are able to perform their function, we should be able to say that ANN can do everything that human neural net can do. In other words, everything that Walter Pitts was hoping for would be possible.

---

[13]The development of this term can be tracked in two general articles written by Creath [10] and Uebel [31].

[14]Maybe it is farfetched, but let us here try to speculate over one more thing. In his paper "Meaning and synonymy in natural languages" [8], Carnap engages the problem of linguistic meaning and the way robots could maybe use it in the future. Although at the time being only a thought experiment, Carnap claimed that knowing their internal structure would help us develop an empirical approach to semantics. In a way, this is similar to Pitts' attempts to completely determine the state of neural networks which would result in possibility to completely describe their behavior.

Remember Leibniz' anticipation of computers, Kant's unknowable thing by itself and the theorem that says there would always exist at least one machine that could do exactly what you say machines cannot do if you were able to completely describe what it is they cannot do? Well, let us use here a common computer phrase that seems suitable yes to all! By mastering logic and by adhering to the rules of logical syntax of language, we would be able to cover both our deductively and inductively gathered knowledge, know the world up to the smallest bits and pieces and precisely and completely describe each action we want. In other words, we would have complete propositional and scientific knowledge needed for describing the world. In this way, we would be able to produce many simple or few complex algorithms that would cover all of our knowledge, leaving nothing unknowable and therefore indescribable. Finally, it would mean every action we can perform would be describable and transferable to ANN, which would now be fully determined as Pitts wanted it to be, opening the possibility of advanced machine learning that would completely correspond to human behavior and learning models. It is safe to say that Carnap influenced Pitts by providing rigorous logical formalism and language syntax offering a way to make the world describable through propositions and scientific, empirical, evidence and, thus, providing an answer to the question that left after the description theorem under which McCulloch's and Pitts' worked. If we consider Turing as paving the way for the development of the research in cognitive frameworks for artificial intelligence in general and (machine) learning in particular, then we can consider McCulloch and Pitts as providing provide a formalization of the physiological part, which grants different insights into the common problem.

In conclusion, let us explicitly state what we hope this chapter could be expected to achieve. Development in different aspects of logic caused the mathematical and philosophical subspecialties to emerge and cause friction between scientists. One would hope to expect to see this friction on the same side of a rift so that it evolves into a fruitful discussion. But alas, the rift occurred in the wrong place—what happened in the last century is not a desirable state of the field:

> In the final analysis, logic deals with reasoning—and relatively little of the reasoning we do is mathematical, while almost all of the mathematical reasoning that nonmathematicians do is mere calculation. To have both rigor and scope, logic needs to keep its mathematical and its philosophical side united in a single discipline. In recent years, neither the mathematical nor the philosophical professions—and this is especially true in the United States—have done a great deal to promote this unity. [28]

It is not a small thing to ask for, but let us hope this chapter is a small step to desired unification, especially with all of philosophical relevance and implications shown in a subject matter that is usually considered to belong to mathematics and computer science.

References

1. Anderson JA, Rosenfeld E (1998) Talking nets: an oral history of neural networks. The MIT Press, Cambridge
2. Arbib MA (2003) The handbook of brain theory and neural networks, 2nd edn. The MIT Press, Cambridge
3. Arbib MA (2019) Warren McCullochs search for the logic of the nervous system. Perspect Biol Med 43(2):193–216
4. Block N (1981) Psychologism and behaviorism. Philos Rev 1(90):5–43
5. Blum M (1989) Notes on McCulloch-Pitts "a logical calculus of the ideas immanent in nervous activity". In: McCulloch R (ed) The collected works of Warren McCulloch, vol 2. Intersystems Publications
6. Bringsjord S, Govindarajulu NS (2018) Artificial intelligence. The Stanford Encyclopedia of Philosophy
7. Buckner C, Garson J (2019) Connectionism. The Stanford Encyclopedia of Philosophy
8. Carnap R (1955) Meaning and synonymy in natural languages. Philos Stud Int J Philo Anal Trad 6(3):33–47
9. Carnap R ([1938], 2001) Logical syntax of language. Routledge, London
10. Creath R (2017) Logical empiricism. The Stanford Encyclopedia of Philosophy
11. De Mol L (2018) Turing machines. The Stanford Encyclopedia of Philosophy
12. Friedman M (1988) Logical truth and analyticity in Carnap's "logical syntax of language". In: Asprey P, Kitcher P (eds) History and philosophy of modern mathematics. University of Minnesota Press, pp 82–94
13. Gefter A (2016) The man who tried to redeem the world with logic. In: Stewart A (ed) The best American science and nature writing 2016. Houghton Mifflin Harcourt, pp 53–65
14. Hilbert D, Ackermann W ([1927], 1959) Grundüge der Theoretischen Logik. Springer, Berlin-Heidelberg
15. Lettvin JL (1989) Introduction. In: McCulloch R (ed) The collected works of Warren McCulloch, vol 1. Intersystems Publications, pp 7–20
16. McCulloch WS, Pitts WH (1943) A logical calculus of the ideas immanent in nervous activity. Bullet Math Biophys 5(4):115–133
17. Oppy G, Dowe D (2019) The turing test. The Stanford Encyclopedia of Philosophy
18. Palm G (1986) Warren McCulloch and Walter Pitts: a logical calculus of ideas immanent in nervous activity. In: Palm G, Aertsen A (eds) Brain theory. Proceedings of the first Trieste meeting on brain theory, October 1–4, 1984. Springer, pp 228–230
19. Piccinini G (2004) The first computational theory of mind and brain: a close look at McCulloch and Pitts's "logical calculus of ideas immanent in nervous activity". Synthese 141(2):175–215
20. Pitts WH (1942) The linear theory of neuron networks: the static problem. Bullet Math Biophys 4(4):169–175
21. Pitts WH (1942) Some observations on the simple neuron circuit. Bullet Math Biophys 4(3):121–129
22. Pitts WH (1943) The linear theory of neuron networks: the dynamic problem. Bullet Math Biophys 5(1):23–31
23. Pitts WH, McCulloch WS (1947) How we know universals: the perception of auditory and visual forms. Bullet Math Biophys 9(3):127–147
24. Rescorla M (2017) The computational theory of mind. The Stanford Encyclopedia of Philosophy
25. Russell S, Norvig P (2002) Artificial intelligence: a modern approach, 2nd edn. Prentice Hall, Saddle River, NJ
26. Searle J (1997) The mystery of consciousness. New York Review of Books, New York
27. Smalheiser NR (2000) Walter pitts. Perspect Biol Med 43(2):217–226
28. Thomason R (2018) Logic and artificial intelligence. The Stanford Encyclopedia of Philosophy
29. Turing A (1936) On computable numbers, with an application to the entscheidungsproblem. Proc Lond Math Soc 42(1):230–265

30. Turing A (1950) Computing machinery and intelligence. Mind 59(236):433–460
31. Uebel T (2019) Vienna circle. The Stanford Encyclopedia of Philosophy
32. Whitehead AN, Russell B (1963) Principia Mathematica. Cambridge University Press, London

Chapter 7
A Lost Croatian Cybernetic Machine Translation Program

Sandro Skansi, Leo Mršić and Ines Skelac

Abstract We are exploring the historical significance of research in the field of machine translation conducted by Bulcsú László, Croatian linguist, who was a pioneer in machine translation in Yugoslavia during the 1950s. We are focused on two important seminal papers written by members of his research group between 1959 and 1962, as well as their legacy in establishing a Croatian machine translation program based around the Faculty of Humanities and Social Sciences of the University of Zagreb in the late 1950s and early 1960s. We are exploring their work in connection with the beginnings of machine translation in the USA and USSR, motivated by the Cold War and the intelligence needs of the period. We also present the approach to machine translation advocated by the Croatian group in Yugoslavia, which is different from the usual logical approaches of the period, and his advocacy of cybernetic methods, which would be adopted as a canon by the mainstream AI community only decades later.

Keywords Bulcsú László · Machine translation · History of technology in Croatia · Language technologies · Natural language processing

The first author's (S.S.) research was supported by the short-term grant *Philosophical Aspects of Logic, Language and Cybernetics* funded by the University of Zagreb under the Short-term Research Support Program.

S. Skansi (✉)
Faculty of Croatian Studies, University of Zagreb, Zagreb, Croatia
e-mail: sskansi@hrstud.hr

L. Mršić · I. Skelac
Algebra University College, Zagreb, Croatia
e-mail: leo.mrsic@algebra.hr

I. Skelac
e-mail: ines.skelac@racunarstvo.hr

S. Skansi (ed.), *Guide to Deep Learning Basics*,
https://doi.org/10.1007/978-3-030-37591-1_7

7.1 Beginnings of Machine Translation and Artificial Intelligence in the USA and USSR

In this chapter, we are exploring the historical significance of Croatian machine translation research group. The group was active in the 1950s, and it was conducted by Bulcsú László, Croatian linguist, who was a pioneer in machine translation during the 1950s in Yugoslavia.

To put the research of the Croatian group in the right context, we have to explore the origin of the idea of machine translation. The idea of machine translation is an old one, and its origin is commonly connected with the work of Rene Descartes, i.e. to his idea of universal language, as described in his letter to Mersenne from 20.xi.1629 [6]. Descartes describes universal language as a simplified version of the language which will serve as an "interlanguage" for translation. That is, if we want to translate from English to Croatian, we will firstly translate from English to an "interlanguage", and then from the "interlanguage" to Croatian. As described later in this chapter, this idea had been implemented in the machine translation process, firstly in the Indonesian-to-Russian machine translation system created by Andreev, Kulagina and Mel'chuk in the early 60s.

In modern times, the idea of machine translation was put forth by the philosopher and logician Yehoshua Bar-Hillel (most notably in [4, 5]), whose papers were studied by the Croatian group. Perhaps the most important unrealized point of contact between machine translation and cybernetics happened in the winter of 1950/51. In that period, Bar-Hillel met Rudolf Carnap in Chicago, who introduced him the (then new) idea of cybernetics. Also, Carnap gave him the contact details of his former teaching assistant, Walter Pitts, who was at that moment with Norbert Wiener at MIT was supposed to introduce him to Wiener, but the meeting never took place [10]. Nevertheless, Bar-Hillel was to stay at MIT where he, inspired by cybernetics, would go to organize the first machine translation conference in the world in 1952 [10].

The idea of machine translation was a tempting idea in the 1950s. The main military interest in machine translation as an intelligence gathering tool (translation of scientific papers, daily press, technical reports and everything the intelligence services could get their hands on) was sparked by the Soviet advance in nuclear technology, and would later be compounded by the success of Vostok 1 (termed by the USA as a "strategic surprise"). In the nuclear age, being able to read and understand what the other side was working on was of crucial importance [25]. Machine translation was quickly absorbed in the program of the Dartmouth Summer Research Project on Artificial Intelligence in 1956 (where artificial intelligence as a field was born), as one of the five core fields of artificial intelligence (later to be known as natural language processing). One other field was included here, the "nerve nets" as they were known back then, today commonly known as artificial neural networks. What is also essential for our discussion is that the earliest programming language for artificial intelligence, Lisp, was invented in 1958 by John McCarthy [18]. But let us take a closer look at the history of machine translation. In the USA, the first major wave of government and military funding for machine translation came in 1954, and the period of abundancy lasted until 1964, when the National Research Council established the Automatic Language Processing Advisory Committee (ALPAC), which was to assess the results of the 10 years of intense funding. The findings were very negative, and funding was almost gone [25], hence the ALPAC report became the catalyst for the first "AI Winter".

One of the first recorded attempts of producing a machine translation system in the USSR was in 1954 [21], and the attempt was applauded by the Communist party of the Soviet Union, by the USSR Committee for Science and Technology and the USSR Academy of Sciences. The source does not specify how this first system worked, but it does delineate that the major figures of machine translation of the time were N. Andreev of the Leningrad State University, O. Kulagina and I. Mel'chuk of the Steklov Mathematical Institute. There is information on an Indonesian-to-Russian machine translation system by Andreev, Kulagina and Mel'chuk from the early 1960s, but it is reported that the system was ultimately a failure, in the same way in which early American systems were a failure. The system had statistical elements set forth by Andreev, but the bulk was logical and knowledge-heavy processing put forth by Kulagina and Mel'chuk. The idea was to have a logical intermediate language, under the working name "Interlingua", which was the connector of both natural languages, and was used to model common-sense human knowledge. For more details, see [21].

In the USSR, there were four major approaches to machine translation in the late 1950s [17]. The first one was the research at the Institute for Precise Mechanics and Computational Technology of the USSR Academy of Sciences. Their approach was mostly experimental and not much different from today's empirical methods. They evaluated the majority of algorithms known at the time and designed their own rather specialized algorithms over meticulously prepared datasets. The main trademarks of their effort were using rather clean data, and by 1959 they have built a German-Russian machine translation prototype. The second approach, as noted by Mulić [17], was championed by the team at the Steklov Mathematical Institute of the USSR Academy of Sciences led by A. A. Reformatsky. Their approach was mainly logical, and they extended the theoretical ideas of Bar-Hillel [5] to build three algorithms: French-Russian, English-Russian and Hungarian-Russian. The third and perhaps the most successful approach was the one by A. A. Lyapunov, O. S. Kulagina and R. L. Dobrushin. Their efforts resulted in the formation of the Mathematical Linguistics Seminar at the Faculty of Philology in Moscow in 1956 and in Leningrad in 1957. Their approach was mainly information-theoretic (but they also tried logic-based approaches [17]), which was considered cybernetic at that time. This was the main role model for the Croatian efforts from 1957 onwards. The fourth, and perhaps most influential, was the logico-statistical approach at the Experimental Laboratory of the Leningrad University championed by N. D. Andreev [17]. Here, the algorithms for Indonesian-Russian, Arabic-Russian, Hindu-Russian, Japanese-Russian, Burmese-Russian, Norwegian-Russian, English-Russian, Spanish-Russian and Turkish-Russian were being built. The main approach of Andreev's group was to use an intermediary language, which would capture the meanings [17]. It was an approach similar to KL-ONE, which would be introduced in the West much later (in

1985) by Brachman and Schmolze [2]. It is also interesting to note that the Andreev group had a profound influence on the Czechoslovakian machine translation program [22], which unfortunately suffered a similar fate as the Croatian program due to the lack of funding.

Andreev's approach was in a sense "external". The modelling would be statistical, but its purpose would not be to mimic the stochasticity of the human thought process, but rather to produce a working machine translation system. Kulagina and Mel'chuk disagreed with this approach as they thought that more of what is presently called "philosophical logic"[1] was needed to model the human thought process at the symbolic level, and according to them, the formalization of the human thought process was a prerequisite for developing a machine translation system (cf. [21]). We could speculate that sub-symbolic processing would have been acceptable too since that approach is also rooted in philosophical logic as a way of formalizing human cognitive functions and is also "internal" in the same sense symbolic approaches are.

There were many other less popular centres for research in machine translation: Gorkovsky University (Omsk), First Moscow Institute for Foreign Languages, Computing Centre of the Armenian SSR and at the Institute for Automatics and Telemechanics of the Georgian SSR [17]. It is worthwhile to note that both the USA and the USSR had access to state-of-the-art computers, and the political support for the production of such systems meant that computers were made available to researchers in machine translation. However, the results were poor in the late 1950s, and a properly working system was yet to be shown. All work done was primarily theoretical work, which was implemented as computer code only after it was completely written and debugged on paper, and this proved to be sub-optimal in the long run.

7.2 The Formation of the Croatian Group in Zagreb

In Yugoslavia, organized effort in machine translation started in 1959, but the first individual effort was made by Vladimir Matković from the Institute for Telecommunications in Zagreb in 1957 in his Ph.D. thesis on entropy in the Croatian language [8]. The main research group in machine translation was formed in 1958, at the Circle for Young Linguists in Zagreb, initiated by the young linguist Bulcsú László, who graduated in Russian language, Southern Slavic languages and English

[1] Which, back then, was somewhat confusingly called "mathematical logic". This difference arises from the fact that up until modal logic took off in the 1960s, the main difference in logic was between "traditional (informal) logic" (or in the case of the Soviet Union it was called "dialectical logic") and the formal version championed by Frege, Russell, Quine and many others which was termed "mathematical logic", to delineate its formal nature. When modal logic semantics came into the picture, an abundancy of philosophical theories could suddenly be formalized, such as time, knowledge, action, duty and paradox and these logics became collectively referred to as "philosophical logic", and the term "mathematical logic" was redefined to include logical topics of interest to mathematicians, such as set theory, recursive structures, algebraically closed fields and topological semantics. This change in terminology was possible since the invention of modal semantics made both fields completely formal.

language and literature at the University of Zagreb in 1952. The majority of the group members came from different departments of the Faculty of Humanities and Social Sciences of the University of Zagreb, with several individuals from other institutions. The members from the Faculty of Humanities and Social Sciences were Svetozar Petrović (Department of Comparative Literature), Stjepan Babić (Department of Serbo-Croatian Language and Literature), Krunoslav Pranjić (Department of Serbo-Croatian Language and Literature), Željko Bujas (Department of English Language and Literature), Malik Mulić (Department of Russian Language and Literature) and Bulcsú László (Department of Comparative Slavistics). The members of the research group from outside the Faculty of Humanities and Social Sciences were Božidar Finka (Institute for Language of the Yugoslav Academy of Sciences and Arts), Vladimir Vranić (Center for Numerical Research of the Yugoslav Academy of Sciences and Arts), Vladimir Matković (Institute for Telecommunications) and Vladimir Muljević (Institute for Regulatory and Signal Devices)[2] [8].

László and Petrović [13] also commented on the state of the art of the time, noting the USA prototype efforts from 1954 and the publication of a collection of research papers in 1955 as well as the USSR efforts starting from 1955 and the UK prototype from 1956. They do not detail or cite the articles they mention. However, the fact that they referred to them in their text published in 1959 (probably prepared for publishing in 1958, based on [13], where Laszlo and Petrović described that the group had started its work in 1958) leads us to the conclusion that the poorly funded Croatian research was lagging only a couple of years behind the research of the superpowers (which invested heavily in this effort). Another interesting moment, which they delineated in [13], is that the group soon discovered that some experimental work had already been done in 1957 at the Institute of Telecommunications (today a part of the Faculty of Electrical Engineering and Computing at the University of Zagreb) by Vladimir Matković. Because of this, they decided to include him in the research group of the Faculty of Humanities and Social Sciences at the University of Zagreb. The work done by Matković was documented in his doctoral dissertation but remained unpublished until 1959.

The Russian machine translation pioneer Andreev expressed hope that the Yugoslav (Croatian) research group could create a prototype, but sadly, due to the lack of federal funding, this never happened [8]. Unlike their colleagues in the USA and the USSR, László's group had to manage without an actual computer (which is painfully obvious in [27]), and the results remained only theoretical. Appealing probably to the political circles of the time, László and Petrović note that, although it sounds strange, research in computational linguistics is mainly a top-priority military effort in other countries [13]. There is a quote from [8] which perhaps best delineates the optimism and energy that the researchers in Zagreb had:

> [...] The process of translation has to mechanicalized as soon as possible, and this is only possible if a competent, fast and inexhaustible machine which could inherit the translation

[2] At the present time, the Institute for Telecommunications and the Institute for Regulatory and Signal Devices are integrated in the Faculty of Electrical Engineering and Computing of the University of Zagreb.

> task is created, even if just schematic. The machine needs to think for us. If machines help humans in physical tasks, why would they not help them in mental tasks with their mechanical memory and automated logic (p. 118).

7.3 Contributions of the Croatian Group

László and Petrović [13] considered cybernetics (used in a broader sense than is usual today, as described in [29] by Wiener) to be the best approach for machine translation in the long run. The question is whether László's idea of using cybernetics would drive the research of the group towards artificial neural networks. László and his group do not go into neural network details (bear in mind that this is 1959–roughly the time around Rosenblatt's research in artificial neural networks), but the following passage offers a strong suggestion about the idea they had (bearing in mind that Wiener relays McCulloch and Pitts' ideas in his book): "Cybernetics is the scientific discipline which studies analogies between machines and living organisms" ([13], p. 107). They fully commit to the idea two pages later ([13], p. 109): "An important analogy is the one between the functioning of the machine and that of the human nervous system". This could be taken to mean a simple computer brain analogy in the spirit of [15] and later [24], but László and Petrović specifically said that thinking of cybernetics to be just the "theory of electronic computers" (as they are made) is wrong [13] since the emphasis of cybernetics should be on modelling computational processes in analogy with human functionality. There is a very interesting quote from [13], where László and Petrović note that "today, there is a significant effort in the world to make a fully automated machine translation possible; to achieve this, logicians and linguists are making efforts on ever more sophisticated problems". This seems to suggest that they were aware of the efforts of logicians (such as Bar-Hillel, and to some degree Pitts, since Wiener specifically mentions logicians-turned-cyberneticists in his book [29]), but still concluded that a wholly cybernetic the approach would probably be a better choice.

László and Petrović [13] argued that, in order to trim the search space, the words would have to be coded so as to retain their information value but to rid the representations of needless redundancies. This was based on previous calculations of language entropy by Matković, and Matković's idea was simple: conduct a statistical analysis to determine the most frequent letters and assign them the shortest binary code. So A would get 101 while F would get 11010011 [13]. Building on that, László suggested that, when making an efficient machine translation system, one has to take into account not just the letter frequencies but also the redundancies of some of the letters in a word [12]. This suggests that the strategy would be as follows: first make a thesaurus, and pick a representative for each meaning, then stem or lemmatize the words, then remove the needless letters from words (i.e. letters that carry little information, such as vowels, but being careful not to equate two different words) and then encode the words in binary strings, using the letter frequencies. After that, the texts are ready for translation, but unfortunately, the translation method is never explicated. Nevertheless, it is hinted that it should be "cybernetic", which, along with what we have presented earlier, would most probably mean artificial neural networks. This is highlighted by the following passage ([13], p. 117):

> A man who spends 50 years in a lively and multifaceted mental activity hears a billion and a half words. For a machine to have an ability comparable to such an intellectual, not just in terms of speed but also in terms of quality, it has to have a memory and a language sense of the same capacity, and for that - which is paramount - it has to have in-built conduits for concept association and the ability to logically reason and verify, in a word, the ability to learn fast.

Unfortunately, this idea of using machine learning (as primitive as it was back in those days) was never fully developed, and the Croatian group regressed back to the Soviet approach(es). Pranjić [23] analyses and extrapolates five basic ideas in the Soviet Machine Translation program, which were the basis for the Croatian approach:

1. Separation of the dictionary from the MT algorithm,
2. Separation of the understanding and generation modules of the MT algorithms,
3. All words need to be lemmatized,
4. The word lemma should be the key of the dictionary, but other forms of the word must be placed as a list in the value next to the key and
5. Use context to determine the meaning of polysemous words.

The dictionary that was mentioned before is, in fact, the intermediary language, and all the necessary knowledge should be placed in this dictionary; the keys should ideally be just abstract codes, and everything else would reside and be accessible as values next to the keys [27]. Petrović, when discussing the translation of poetry [20], noted that ideally, machine translation should be from one language to another, without the use of an intermediate language of meanings.

Finka and Lászlо́ envisioned three main data preparation tasks that are needed before prototype development could commence [8]. The first task is to compile a dictionary of words sorted from the end of the word to the beginning. This would enable the development of what is now called stemming and lemmatization modules: not only a knowledge base with suffixes so they can be trimmed but also a systematic way to find the base of the word (lemmatization) (p. 121). The second task would be to make a word frequency table. This would enable focusing on a few thousand most frequent words and dropping the rest. This is currently a good industrial practice for building efficient natural language processing systems, and in 1962, it was a computational necessity. The last task was to create a good thesaurus, but such a thesaurus where every data point has a "meaning" as the key, and words (synonyms) as values. The prototype would then operate on these meanings when they become substituted for words.

But what are those meanings? The algorithm to be used was a simple statistical alignment algorithm (in hopes of capturing semantics) described in [27] on a short Croatian sentence "čovjek [noun-subject] puši [verb-predicate] lulu [noun-objective]" (A man is smoking a pipe). The first step would be to parse and lemmatize. Nouns in Croatian have seven cases just in the singular, with different suffixes, for example:

ČOVJEK - Nominative singular
ČOVJEKA - Genitive singular
ČOVJEKU - Dative singular
ČOVJEKA - Accusative singular
ČOVJEČE - Vocative singular
ČOVJEKU - Locative singular
ČOVJEKOM - Instrumental singular

Although morphologically transparent, the lemma in the mentioned case would be "ČOVJEK-"; there is a voice change in the Vocative case, so for the purpose of translation, "ČOVJE-" would be the "lemma". The other two lemmas are PUš- and LUL-.

The thesaurus would have multiple entries for each lemma, and they would be ordered by descending frequency (if the group actually made a prototype, they would have realized that this simple frequency count was not enough to avoid only the first meaning to be used). The dictionary entry for ČOVJE- (using modern JSON notation) is

"ČOVJE-": "mankind": 193.5: "LITTLENESS", 690.2: "AGENT", "man": 554.4: "REPRESENTATION", 372.1: "MANKIND", 372.3: "MANKIND" ..., ...

The meaning of the numbers used is never explained, but they would probably be used for cross-referencing word categories.

After all the lemmas comprising the sentence have been looked up in this dictionary, the next step is to keep only the inner values and discard the inner keys, thus collapsing the list so that the example above would become

"COVJE-": 193.5: "LITTLENESS", 690.2: "AGENT", 554.4: "REPRESENTATION", 372.1: "MANKIND", 372.3: "MANKIND" ...

Next, the most frequently occurring meaning would be kept, but only if it grammatically fits the final sentence. One can extrapolate that it is tacitly assumed that the grammatical structure of the source language matches the target language, and to do this, a kind of categorical grammar similar to Lambek calculus [11] would have to be used. It seems that the Croatian group was not aware of the paper by Lambek (but only of Bar-Hillel's papers), so they did not elaborate on this part.

Finka [7] notes that Matković, in his dissertation from 1957, considered the use of bigrams and trigrams to "help model the word context". It is not clear whether Finka means character bigrams, which was computationally feasible at the time, or word bigrams, which was not feasible, but the suggestion of modelling the word context does point in this direction. Even though the beginnings of using character bigrams can be traced back to Claude Shannon [26], using character-level bigrams in natural language processing was studied extensively only by Gilbert and Moore [9]. It can be argued, that in a sense, Matković predated these results, but his research and ideas were not known in the West, and he was not cited. The successful use of word bigrams in text classification had to wait until [14]. The long time it took to get from character to words was mainly due to computational limitations, but Matković's ideas are not to be dismissed lightly on account of computational complexity since the idea of using word bigrams was being explored by the Croatian group—perhaps the reason

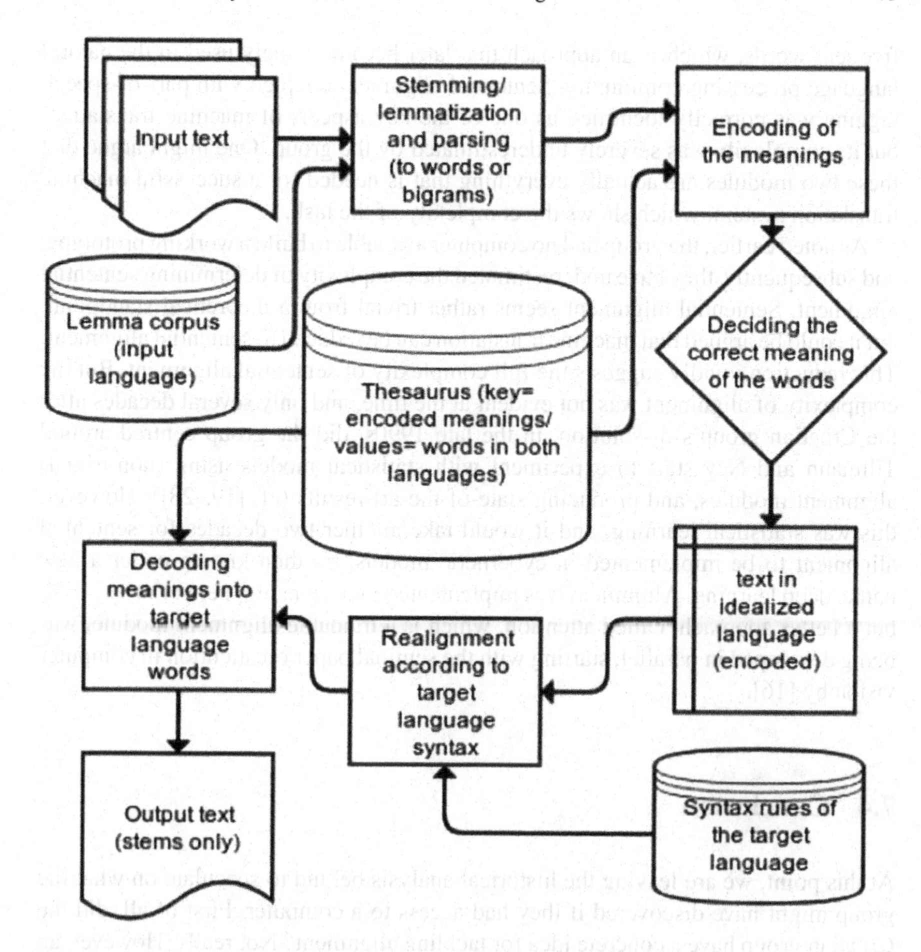

Fig. 7.1 A reconstruction of the Croatian group's prototype

for considering such an idea was the lack of a computer and the underestimation of the memory requirements. The whole process described above is illustrated in Fig. 7.1.

Several remarks are in order. First, the group seemed to think that encodings would be needed, but it seems that entropy-based encodings and calculations added no real benefits (i.e. added no benefit that would not be offset by the cost of calculating the codes). In addition, Finka and László [8] seem to place great emphasis on lemmatization instead of stemming, which, if they had constructed a prototype, they would have noticed it to be very hard to tackle with the technology of the age. Nevertheless, the idea of proper lemmatization would probably be replaced with moderately precise hard-coded stemming, made with the help of the "inverse dictionary", which Finka and László proposed as one of the key tasks in their 1962 paper. This paper also highlights the need for a frequency count and taking only the most frequent words, which is an approach that later became widely used in the natural language processing community. Sentential alignment coupled with part-of-speech tagging was correctly identified as one of the key aspects of machine translation, but its complexity was severely underestimated by the group. One might argue that these two modules are actually everything that is needed for a successful machine translation system, which shows the complexity of the task.

As noted earlier, the group had no computer available to build a working prototype, and subsequently, they have underestimated the complexity of determining sentential alignment. Sentential alignment seems rather trivial from a theoretical standpoint, but it could be argued that machine translation can be reduced to sentential alignment. This reduction vividly suggests the full complexity of sentential alignment. But the complexity of alignment was not evident at the time, and only several decades after the Croatian group's dissolution, in the late 1990s, did the group centred around Tillmann and Ney start to experiment with statistical models using (non-trivial) alignment modules, and producing state-of-the-art results (cf. [19, 28]). However, this was statistical learning, and it would take another two decades for sentential alignment to be implemented in cybernetic models, by then known under a new name, deep learning. Alignment was implemented in deep neural networks by [1, 3], but a better approach, called attention, which is a trainable alignment module, was being developed in parallel, starting with the seminal paper on attention in computer vision by [16].

7.4 Conclusion

At this point, we are leaving the historical analysis behind to speculate on what the group might have discovered if they had access to a computer. First of all, did the Croatian group have a concrete idea for tackling alignment? Not really. However, an approach can be read between the lines of primarily [12, 23]. In [23], Pranić addresses the Soviet model by Andreev, looking at it as if it was composed of two modules—an understanding module and a generation module. Following the footsteps of Andreev, their interaction should be over an idealized language. László [12] notes that such an idealized language should be encoded by keeping the entropy in mind. He literally calls for using entropy to eliminate redundancy while translating to an artificial language, and as Mulić notes [17], Andreev's idea (which should be followed) was to use an artificial language as an intermediary language, which has all the essential structures of all the languages one wishes to translate.

The step which was needed here was to eliminate the notion of structure alignment and just seek sentential alignment. This, in theory, it can be done by using only entropy. A simple alignment could be made by using word entropies in both languages and aligning the words by decreasing entropy. This would work better for translating into a language with no articles such as Russian or Croatian. A better approach, which was not beyond the thinking of the group since it was already proposed by Matković in his dissertation from 1957 [7], would be to use word bigrams and align them. It is worth mentioning that, although the idea of machine translation in the 1950s in Croatia did not have a significant influence on the development of the field, it shows that Croatian linguists had contemporary views and necessary competencies for its development. But, unfortunately, the development of machine translation in Croatia had been stopped because of the previously discussed circumstances. In 1964, László went to the USA, where he spent the next 7 years, and after returning to Croatia, he was active as a university professor, but because of disagreement with the ruling political option regarding Croatian language issues, he published very rarely and was mainly focused on other linguistic issues in that period, but his work was a major influence on the later development of computational linguistics in Croatia.

References

1. Alkhouli T, Bretschner G, Peter JT, Hethnawi M, Guta A, Ney H (2016) Alignment-based neural machine translation. In: Bojar O, Buc Ch et al (eds) Proceedings of the first conference on a machine translation, volume 1: Research papers, WMT 2016, Berlin, Germany, 7–12 Aug 2016. Association for Computational Linguistics, Berlin, pp 54–65
2. Baader F, Horrocks I, Sattler U (2007) Description logics. In: van Harmelen F, Lifschitz V, Porter B (eds) Handbook of knowledge representation. Elsevier, Oxford, pp 135–180
3. Bahbanau D, Cho KH, Bengio Y (2015) Neural machine translation by jointly learning to align and translate. In: International conference on learning representations, ICLR 2015, San Diego, CA
4. Bar-Hillel Y (1951) The present state of research on mechanical translation. Am Doc 2:229–236
5. Bar-Hillel Y (1953) Machine translation. Comput Autom 2:1–6
6. Descartes R (2017) Selected correspondence of descartes. Early Modern Philosophy, Cambridge
7. Finka B (1959) Odjeci sp u jugoslaviji. In: Laszlo B, Petrovic S (eds) Strojno prevodenje i statistika u jeziku. Zagreb, Naše teme, pp 249–260
8. Finka B, Laszlo B (1962) Strojno prevodenje i naši neposredni zadaci [machine translation and our immediate tasks]. Jezik 10:117–121
9. Gilbert EN, Moore EF (1959) Variable-length binary encodings. Bell Syst Tech J 38:933–967
10. Hutchins J (2000) Yehoshua bar-hillel. a philosopher's contribution to machine translation. In: Hutchins J (ed) Early years in machine translation. John Benjamins, Amsterdam, pp 299–312
11. Lambek J (1958) The mathematics of sentence structure. Am Math Mon 65:154–170
12. Laszlo B (1959) Broj u jeziku. In: Laszlo B, Petrovic S (eds) Strojno prevodenje i statistika u jeziku. Zagreb, Naše teme, pp 224–239
13. Laszlo B, Petrovic S (1959) Uvod. In: Laszlo B, Petrovic S (eds) Strojno prevodenje i statistika u jeziku. Naše teme, Zagreb, pp 105–298
14. Lewis DD (1992) An evaluation of phrasal and clustered representations on a text categorization task. In: Belkin NJ, Ingwersen P, Pejtersen AM (eds) Proceedings of SIGIR '92, 15th ACM international conference on research and development in information retrieval, SIGIR '92, Copenhagen, Denmark, 21–24 June 1992. ACM Press, New York, pp 37–50
15. McCulloch W, Pitts W (1943) A logical calculus of ideas immanent in nervous activity. Bull Math Biophys 5:115–133
16. Mnih V, Heess N, Graves A, Kavukcuoglu K (2014) Recurrent models of visual attention. In: Ghahramani Z et al (eds) NIPS'14 Proceedings of the 27th international conference on neural information processing systems - volume 2, NIPS 2014, Montreal, Canada, 08–13 Dec 2014. MIT Press, Cambridge, MA, pp 2204–2212

17. Mulic M (1959) Sp u sssr. In: Laszlo B, Petrovic S (eds) Strojno prevodenje i statistika u jeziku [Machine translation and statistics in language]. Zagreb, Naše teme, pp 213–221
18. Norvig P (1991) Paradigms of artificial intelligence programming. Morgan Kaufmann, San Francisco
19. Och FJ, Tillmann Ch, Ney H (1999) Improved alignment models for statistical machine translation. In: Schiitze H, Su K-Y (eds) Joint SIGDAT conference on empirical methods in natural language processing and very large corpora, SIGDAT 1999, College Park, MD, USA, 21–22 June 1999. Association for Computational Linguistics, Stroudsburg, PA, pp 20–28
20. Petrovic S (1959) Može li stroj prevoditi poeziju. In: Laszlo B, Petrovic S (eds) Strojno prevodenje i statistika u jeziku. Zagreb, Naše teme, pp 177–197
21. Piotrowski R, Romanov Y (1999) Machine translation in the former soviet union and in the newly independent states. Hist Epistem Lang 21:105–117
22. Pogbrelec B (1959) Počeci rada na sp. In: Laszlo B, Petrovic S (eds) Strojno prevodenje i statistika u jeziku. Zagreb, Naše teme, pp 222–223
23. Pranjic K (1959) Suvremeno stanje sp. In: Laszlo B, Petrovic S (eds) Strojno prevodenje i statistika u jeziku. Zagreb, Naše teme, pp 224–239
24. Putnam H (1979) Mathematics, matter, and method: philosophical papers, vol 1. The MIT Press, Cambridge, MA
25. Russell S, Norvig P (2009) Artificial intelligence, a modern approach. Pearsons, New York
26. Shannon CE (1948) A mathematical theory of communication. Bell Syst Tech J 27:379–423
27. Spalatin L (1959) Rječnik sinonima kao jezik posrednik. In: Laszlo B, Petrovic S (eds) Strojno prevodenje i statistika u jeziku. Zagreb, Naše teme, pp 240–248
28. Vogel S, Ney SH, Tillmann HC (1996) Hmm-based word alignment in statistical translation. In: Tsujii J (ed) 16th international conference on computational linguistics, proceedings of the conference, COLING '96, Copenhagen, Denmark, 05–09 Aug 1996, Center for Sprogteknologi, Copenhagen, pp 836–841
29. Wiener N (1948) Cybernetics: on control and communication in the animal and the machine. The MIT Press, Cambridge, MA

Chapter 8
The Architectures of Geoffrey Hinton

Ivana Stanko

Abstract Geoffrey Everest Hinton is a pioneer of deep learning, an approach to machine learning which allows computational models that are composed of multiple processing layers to learn representations of data with multiple levels of abstraction, whose numerous theoretical and empirical contributions have earned him the title the Godfather of deep learning. This chapter offers a brief outline of his education, early influences and prolific scientific career that started in the midst of AI winter when neural networks were regarded with deep suspicion. With a single goal fueling his ambitions—understanding how the mind works by building machine learning models inspired by it—he surrounded himself with like-minded collaborators and worked on inventing and improving many of the deep learning building blocks such as distributed representations, Boltzmann machines, backpropagation, variational learning, contrastive divergence, deep belief networks, dropout, and rectified linear units. The current deep learning renaissance is the result of that. His work is far from finished; a revolutionary at heart, he is still questioning the basics and currently developing a new approach to deep learning in the form of capsule networks.

Keywords Geoffrey Hinton · Restricted Boltzmann machines · Cognitive science · Distributed representations, Backpropagation algorithm, Deep learning

8.1 Context

The field of artificial intelligence has been divided into two opposing camps on the converging paths toward the ultimate quest of "making machines do things that would require intelligence if done by men"[1]—those who want to *program* machines to do things (knowledge-based approach based on manipulating symbols according to the rules of logic) and those who want to *show* machines how to do things (machine

[1] Marvin Minsky.

I. Stanko (✉)
University of Zagreb, Zagreb, Croatia
e-mail: istanko@hrstud.hr

S. Skansi (ed.), *Guide to Deep Learning Basics*,
https://doi.org/10.1007/978-3-030-37591-1_8

learning based on applied statistics and extracting patterns from raw data). After many decades and many a hard-won battle, the winner is clear-cut and accumulated efforts of generation upon generation of great thinkers from many diverse fields crystallized in the approach to machine learning called deep learning which allows computational models that are composed of multiple processing layers to learn representations of data with multiple levels of abstraction [25]. Credit assignment as the central problem of machine learning aside, it is hard to dispute that the final steps toward the current deep learning renaissance have been spearheaded by the unshakable and deep belief of Geoffrey Hinton, the Godfather of deep learning, and the person behind many of its principal components.

Geoffrey Everest[2] Hinton, an Emeritus Distinguished Professor at the University of Toronto, a Vice President and Engineering Fellow at Google, Chief Scientific Adviser of the Vector Institute, an ACM Turing Award Laureate and one of the Deep Learning Conspiracy[3] trio along with Yoshua Bengio and Yann LeCun, was born on December 6, 1947 in London to a family of distinguished and accomplished individuals. In the context of Hinton's life endeavor, the most prominent node in his family tree represents George Boole whose seminal work *An Investigation of the Laws of Thought* introduced Boolean algebra that has not only been instrumental in the development of digital computing but also offered many less-remembered insights into probability theory, the bedrock of machine learning, that have come to fruition through the research of his great-great-grandson [36]. Hinton's mother was a school teacher who made it clear early that he had choices in life—he could either be an academic (like his father, an entomologist) or a failure[4]—so by the age of 7, he already knew that getting a Ph.D. was nonnegotiable.[5] His father, a Stalinist, sent him to a private Catholic school, an education which in addition to a regular curriculum included mandatory morning prayers, which taught him, among other things, that people's beliefs are often nonsense [31]; a most useful training example that enabled him to generalize well throughout his professional career.

School, in general, was not a particularly happy time for Hinton; he spent the whole 11 years feeling like an outsider which lit up a contrarian spark in him and prepared him for the revolutionary role he would assume in the development of neural networks [31]. It was in high school that his interest took shape during illuminating conversations with his friend Inman Harvey who introduced Hinton to the idea that human memory might work as a hologram, a clear-cut example of distributed representations, where each memory is stored by adjusting connection strengths between the neurons across the entire brain [4]. His first foray into university ended with him dropping out after a month and doing odd jobs here and there while his life goals

[2]Name shared with a relative George Everest, a surveyor and a geographer, after whom Mount Everest was named.

[3]Another popular nickname is Canadian Mafia.

[4]*The Godfather of Deep Learning was Almost a Carpenter*. For full interview see https://www.bloomberg.com/news/videos/2017-12-01/the-godfather-of-ai-was-almost-a-carpenter-video.

[5]*This Canadian Genius Created Modern AI*. For the full interview see https://www.youtube.com/watch?v=l9RWTMNnvi4.

crystallized [31]. The second time around, he started studying Physics and Physiology; Physiology did not even try to explain the how behind the mechanics of the brain function and physics left him drowning in a sea of equations he considered too difficult so he switched to Philosophy for a year hoping to find the answer to the meaning of life—he did not and was left unsatisfied because the framework of the discipline offered no way to verify whether you were right or wrong [31]. the third time was not the charm and his next stop, Psychology, turned out not to be the royal road to understanding the complexities of the mind—the emphasis was on rats in the mazes, models of how the mind worked were hopelessly inadequate and it was only after Hinton spurred other students to protest the content of the course that the department organized a single perfunctory lecture on Freud, Jung, and Adler—his discontent with the entire field continued to grow [31]. Ditching the academia altogether, he tried out the life of a carpenter but having met an expert one Hinton realized his inadequacies [4] and returned to Cambridge where he earned his B. A. Hons in Experimental Psychology in 1970.

Having realized that the only way to truly understand a complex device such as the human brain is to build one,[4] he set his sights on artificial intelligence and went to Edinburgh University where he began his graduate studies under the supervision of Christopher Longuet-Higgins, a renowned chemist who coined the term *cognitive science* and envisioned it as a unification of mathematical, linguistic, psychological, and physiological sciences, with AI as the key ingredient that would elucidate how the human mind worked [26]. It was the shared view on AI's real purpose and Longuet-Higgins' work on an early model of associative memory that made Hinton choose him as his thesis advisor. Unfortunately, just before Hinton arrived Longuet-Higgins' interest in neural networks had started to wane as he found himself impressed with Terry Winograd's dissertation *Procedures as a Representation for Data in a Computer Program for Understanding Natural Language* and its star SHRDLU. Longuet-Higgins assumed Hinton wanted to follow the same path and work on symbolic AI; many fights ensued but convinced that neural networks were the only way to go for what Hinton persisted in doing what he believed in.[6] Despite the disagreements and a lack of belief in Hinton's ideas, Longuet-Higgins still supported his right to pursue them and provided valuable advice about building everything on solid mathematical foundations [31]. Hinton was awarded his Ph.D. in 1978 but the prevailing attitude in Britain was that neural networks were a complete waste of time which prevented Hinton not only from finding a job but also from landing a single job interview; a completely different atmosphere awaited at the University of California where he spent time collaborating with Don Norman and David Rumelhart in a fertile intellectual environment that encouraged the to and fro of artificial neural networks and Psychology.[6]

[6] *Heroes of deep learning: Andrew Ng interviews Geoffrey Hinton*. For full interview see https://www.deeplearning.ai/blog/hodl-geoffrey-hinton/.

8.2 Building Blocks

Psychology had long questioned the localizationist doctrine—one computational element for one entity—and the work of Jackson, Luria, Lashley, Hebb, Rosenblatt, and Selfridge, among many others, pointed toward a different way of implementing propositional knowledge in the processing system like the human brain. It seemed more likely that concepts were implemented by **distributed representations** where each entity is represented by a pattern of activity distributed over many computing elements, and each computing element is involved in representing many different entities [15], which automatically enabled generalization because similar objects are represented by similar patterns and property-inheritance whenever the units that are active in the representation of a type are also a subset of the units active in the representation of an instance of that type [6]. Whatever the brain was doing obviously worked and following Orgel's Second Rule "Evolution is cleverer than you are", Hinton knew that distributed representations provided a promising model of implementing semantic networks in parallel hardware. In order to use them effectively the learning problem of finding a good pattern of activity, i.e., the underlying complex regularities, needed to be solved.

An early deep learning model that attempted to solve that problem appeared in 1983 under an intentionally veiled article title *Optimal perceptual inference*, a necessary decision given that even mentioning neural networks was frowned upon in most academic circles. In it Hinton and Terrence Sejnowski hinted at **Boltzmann machines**—energy-based models, which are essentially stochastic Hopfield networks with symmetrically connected hidden[7] units, they developed during Hinton's time at Carnegie Mellon. The energy minimum of a Boltzmann machine corresponds to a distributed representation and the whole system creates a good collection of them by clamping some of the individual units into certain states that represent a particular input, which allows it to find a good energy landscape that is compatible with that input [1]. This is achieved with an unsupervised learning algorithm that requires only local information and uses hidden units to capture the best regularities in the environment by minimizing the Kullback–Leibler divergence by changing the weights in proportion to the difference between the data-dependent expected value of the product of states at thermal equilibrium and the corresponding data-independent expected value [19]. The problem was that the learning itself was slow, noisy, and impractical; Sejnowski believed it could be vastly improved if there were a way to learn smaller modules independently [12]. He was right, but it would take almost 20 years and additional tinkering to transform Boltzmann machines into an essential component of many deep probabilistic models they are today.

Another seed that would require time to come into full bloom and signal the arrival of the AI spring was planted in 1986 when Rumelhart, Hinton, and Ronald Williams managed to get the *Learning representations by back-propagating errors* paper that has since then become classic published in Nature. That itself required some political work[6]—realizing its importance for the progress in the field of AI, Hinton talked to

[7]The name for those units was actually inspired by Hidden Markov Models [31].

one of the suspected referees Stuart Sutherland, a well-known British psychologist, and explained to him the value of using a relatively simple learning procedure to learn distributed representations which, fortunately, did not fall on deaf ears. **Backpropagation**, or simply backprop, uses the chain rule of calculus to compute the gradient allowing another algorithm, like stochastic gradient descent, to perform the learning [5]. Variations of the algorithm date back to 1960s and the work of Henry J. Kelly, Arthur E. Bryson, and Stuart Dreyfus while in the 1970s, Seppo Linnainmaa and Paul Werbos worked on implementing it. When Rumelhart rediscovered it in 1981 and fine-tuned it with Hinton and Williams, nothing spectacular happened; only after returning to it after his work on Boltzmann machines did Hinton realize its full potential and the trio resumed their work on it, applying backpropagation to word embeddings to learn distributed representations and seeing semantic features emerge from it [4].

The network was given information expressed in sets of triplets *person1-relationship-person2* in a graph structure, a family tree task, and succeeded in converting that information into big feature vectors from which emerged meanings of individual features like the nationalities, generations, and branches of the family tree that represented people, allowing the network to derive new information that was used to predict the third item given the first two [32]. Backpropagation took the network from graph structure (input) to features and their interactions (hidden units) and back again (output) by repeatedly adjusting the connection strengths in the network (weights) in a backward pass, starting from the penultimate layer and making its way down the earlier ones to minimize the total error between the actual and desired output [32]. Equally important was the fact that backpropagation rose up to the challenge presented in Minsky's and Papert's 1969 *Perceptrons* and provided solutions to the problems with Rosenblatt's invention—multilayer neural networks were now able to handle abstract mathematical problems such as the XOR and other parity problems, encoding problems, binary addition, and negation as well as mirror symmetry problems and geometric problems such as discrimination between T and C independent of their translation and rotation [33].

Together with Alexander Waibel, Toshiyuki Hanazawa, Kiyohiro Shikano, and Kevin Lang, Hinton made his first practical application by developing a version of backpropagation called time-delay neural networks, which was used for successful phoneme recognition [38]. A researcher at heart, his interest did not lie in developing applications; they were a means to an end—proving something was useful enough to keep the funding flowing while his sights were set on figuring out how the brain works [31]. Despite it being a major breakthrough that started to turn the tide within the AI community, backpropagation did not work, as it was expected, well at the time. Today, when it is used in probably 90% of commercial and industrial applications of neural networks [28], it is relatively easy to see what held it back in the 80s and 90s. In Hinton's own words, the labeled datasets that were used were thousands of times too small, the computers were millions of times too slow, the weights were initialized in a stupid way, and a wrong type of nonlinearity was used.[8] Time would take care of the first two, while the joint effort of Hinton and his collaborators would be behind rectifying the rest.

8.3 Tinkering

As the temporary hype died out so did the funding. Left with the option of accepting military (Office of Naval Research) money to continue his work and generally dissatisfied with the politics that prevailed in American society, Hinton moved to Canada where he took up a position at the Department of Computer Science at the University of Toronto and was able to continue his research minimally interrupted by teaching obligations—thanks to the funding he received from the Canadian Institute for Advanced Research (CIFAR) [31]. His aspiration was to foster the transfer of ideas between the recent advances in neural networks and methods used in statistics, mainly variational methods that included deterministic approximation procedures that generally provided bounds on probabilities of interest [22].

One of the popular ones was the **expectation maximization** (EM) algorithm—a method for tackling approximate inference problems in undirected graphical models with visible and latent variables where computing the log probability of observed data is too difficult so a lower bound (the negative variational free energy) is computed instead and the entire inference problem boils down to finding an arbitrary probability distribution over latent variables that maximizes the lower bound [5]. The EM algorithm alternates between an expectation (E) step that finds the distribution of the latent variables given the known values of the visible ones and the current estimate of the parameters and a maximization (M) step that adjusts the parameters according to maximum likelihood, assuming that the distribution from the E step is correct [30]. Radford Neal and Hinton improved it by making a generalization of it which can be seen in terms of Kullback–Leibler divergence by showing that there was no need to perform a perfect E step, an approximation of it computed by recalculating the distribution for only one of the latent variables would suffice and would converge faster to a solution in mixture estimation problems [30].

His attention turned to unsupervised learning and **autoencoders**, neural networks trained to first convert an input vector into a code vector using recognition weights and then to convert that code vector into an approximate reconstruction of an input using generative weights [21]. Unsupervised learning is a more plausible model of the way humans learn and can be implemented in machine learning by minimizing the sum of a code cost, the number of bits necessary to describe the activities of the hidden units, a reconstruction cost, the number of bits necessary to describe the difference between the input, and its best approximation reconstructed by the activities

[8]Geoffrey Hinton's keynote speech at the 2015 Royal Society *Machine learning: breakthrough science and technologies—Transforming our future conference series*. For full lecture see https://www.youtube.com/watch?v=izrG86jycck/.

of the hidden units [7]. Working with Richard Zemel, they devised a way of training autoencoders using a variation of Minimum Description Length principle by using nonequilibrium Helmholtz free energy as an objective function that simultaneously minimizes the information provided by the activities of the hidden units and the information contained in the reconstruction error [21].

Peter Dayan and Radford Neal joined them and together they generalized that to a multilayer system—**the Helmholtz machine** which is a neural network consisting of multiple layers of binary stochastic units that are connected hierarchically by two sets of weights—bottom-up implement a recognition model that infers a probability distribution over hidden variables given the input, while top-down enforce a generative model that reconstructs the values of the input from the activities of the hidden units [3]. Being an unsupervised neural network, there is no external teaching signal to match so the hidden units need to extract the underlying regularities from the data by learning representations that are economical in their description length but sufficient for an accurate reconstruction of the data, which can be achieved with the **wake-sleep algorithm** [13]. In the wake phase, the bottom-up recognition weights produce a representation of the input in each layer and combine them to determine a conditional probability distribution over total representations; the top-down generative weights are then modified using the delta rule to become better at reconstructing the activities in preceding layers [13]. The sleep phase is driven by top-down generative weights that provide an unbiased sample of the network generative model that is used to train the bottom-up connections again using the delta rule [13].

Another technique Hinton developed for training undirected graphical models was **contrastive divergence** first used on **products of experts** which model high-dimensional data by multiplying and renormalizing several probability distributions of different low-dimensional constraints of that data [8]. Features were learned by following the approximation of the gradient of contrastive divergence, an objective function that removes the mistakes the model generates by minimizing the difference between the two Kullback–Leibler divergences—the one between the data distribution and the equilibrium distribution over the visible variables produced by the Gibbs sampling from the generative model and the other between the reconstructions of the data vectors from the last seen example and the equilibrium distribution—which is equivalent to maximizing the log likelihood of the data [9]. This technique could be applied to **Restricted Boltzmann machines** (RBMs), undirected probabilistic graphical models containing a layer of observable and a layer of latent variables [5], which could be viewed as products of experts with one expert per hidden unit [9]. The learning procedure became much simpler than in the original Boltzmann machines, thus making RBMs practical enough to become not only a staple in the construction of the upcoming deep architectures but also powerful enough on their own to become a part of a million-dollar winning entry to Netflix collaborative filtering competition [35].

8.4 Deep Learning

In 2006 neural networks reemerged under a new name—*deep learning*—and a new era began ushered in by Hinton's, Simon Osindero's and Yee-Whye Teh's, now legendary, paper *A fast learning algorithm for Deep Belief Nets*. In it, they demonstrated that multilayered networks could not only be trained efficiently but could also outperform the models that had dominated the machine learning landscape. This was accomplished by using greedy layer-wise pretraining that corrected the earlier mistake of initializing the weights stupidly—the structure in the input was no longer ignored as with backpropagation alone because by starting off with unsupervised learning, the network could discover latent variables (features) that captured the structure in the training data allowing discriminative learning that followed to model the dependence of the output on the input by fine-tuning those discovered features to discriminate better [11].

The **deep belief network** (DBN) they presented was a multilayer stack of RBMs, the top two hidden layers formed an undirected associative memory and the remaining hidden layers received directed top-down connections and converted the representations in the associative memory into observable variables [16]. By using contrastive divergence learning in RBMs to learn one layer of features at a time, each layer modeling the previous one, the whole learning process was broken down into a sequence of simpler tasks avoiding the inference problems that otherwise appeared in directed generative models [10]. The more layers the network had, the better it performed by exploiting the fact that natural signals are compositional hierarchies [25] and each new hidden layer became an improvement on the variational bound on the log probability of the training data [10]. Generative abilities of the network could be improved by using a contrastive version of the wake-sleep algorithm to fine-tune the learned weights [16], whereas backpropagation could be used to improve the performance on discrimination tasks [18].

The area that was transformed first by these innovations was speech recognition, formally dominated by Hidden Markov Models. In 2009 DBNs used for acoustic modeling by Abdel-rahman Mohamed, George Dahl, and Hinton outperformed other state-of-the-art models on TIMIT [27]. The technology was then offered to RIM,[9] which declined but Google found it interesting enough and by 2012 it was implemented in Android.[8] The academia finally accepted neural networks, at least those who did not subscribe to Max Planck's view on the progress of science,[10] the industry started to take notice but it took some additional tweaking for deep learning to establish itself as the dominant approach to AI.

The problem of overfitting needed to be solved and that was accomplished by taking a leaf out of evolution's book, the chapter on sex, which gave birth to the idea of **dropout**. Sexual reproduction breaks-up sets of complicated coadapted genes, achieving robustness in functionality by forcing each to pull its own weight and mix well with random ones rather than being useful only in tandem with a large

---

[9]Research in Motion, known as BlackBerry Limited since 2013.

[10]"Science advances one funeral at a time."

number of others already present, thus reducing the probability that small changes in the environment will lead to large decreases in fitness [20]. Dropout, developed by Nitish Srivastava, Alex Krizhevsky, Ilya Sutskever, Ruslan Salakhutdinov, and Hinton, simulates ensemble learning and leads to similar robustness by adding noise to the states of hidden units in a neural network [37]. The noise temporarily removes a fixed fraction of the feature detectors and their connections on each presentation of each training case, preventing units from co-adapting too much and forcing individual neurons to learn to detect generally helpful features for arriving at the correct solution given the combinatorially large variety of internal contexts in which they must operate [20]. As a result, dropout has improved generalization performance on tasks across many different domains such as vision, speech recognition, document classification, and computational biology with the only trade-off being increased training time [37].

A simple and effective solution to that problem, and the longstanding issue of using the wrong kind of nonlinearity, was in the synergistic effects of using dropout with a special kind of activation function: **Rectified Linear Unit** (ReLU) [2] that performs better than the logistic sigmoid function and would become the default choice in deep neural networks. Thought of by three different groups of researches, ReLUs have become the core of every deep network where they sum the weighted input of a unit into its activation or its output according to g(x) = $\max\{0, x\}$. It is a piecewise linear function with two linear pieces and applying it to linear transformation gives a nonlinear transformation that outputs zero across half its domain which makes its derivatives large whenever it is active and gradients consistent; this mimics biological neurons and their sparse and selective activations that depend on the input [5]. Vinod Nair and Hinton used it in RBMs as an approximation to an infinite set of replicated binary units with tied weights and shifted biases so as to maintain the RBMs probabilistic model and it worked better than binary hidden units for recognizing objects and comparing faces [29]. When combined with dropout, ReLUs no longer overfit quickly in comparison to sigmoid nets and they worked well together leading to improvements in error reduction over other deep neural network models on LVCSR [2].

These two new insights were combined with the efficient use of GPUs and data augmentation, and then applied to a well-developed albeit overlooked deep architecture called a convolutional neural network and **AlexNet** was conceived. Its landslide victory on the ImageNet Large-Scale Visual Recognition Challenge (ILSVRC) was the tipping point that finally transformed deep learning from a niche interest to an overnight sensation. The competition's goal was to get the lowest top-1 and top-5 error rates in the detection and correct classification of objects and scenes belonging to 1000 different categories with 1000 images in each, totaling approximately 1.2 million training, 50,000 validation and 150,000 testing images. Krizhevsky, Sutskever, and Hinton accomplished exactly that in 2012 and explained how in the seminal *ImageNet Classification with Deep Convolutional Neural Networks* paper.

AlexNet consisted of eight learned layers—the first five were convolutional and the other three fully connected; the output of each was applied to the ReLU allowing much faster learning than with saturating neurons while the output of the last fully connected layer was fed to a 1000-way softmax which produced a distribution over the 1000 class labels [24]. The kernels of the second, fourth, and fifth convolutional layers were connected only to those kernel maps in the previous layer which resided on the same GPU, allowing for an additional trick—GPU communication only in certain layers—along with the parallelization [24]. Response-normalization layers, which aid generalization by creating competition for big activities among neuron outputs that are computed using different kernels thus implementing a variation of lateral inhibition inspired by the one biological neurons use, followed the first and second convolutional layers [24]. Since the architecture contained 60 million parameters, in order to reduce overfitting dropout was used in the first two fully connected layers and the dataset itself was artificially enlarged by generating more training examples by deforming existing ones with image translations, horizontal reflections and altering the intensities of the RGB channels [24]. They achieved the error rate of 15.3%, the runner-up's was 26.2% and it was the depth of AlexNet that enabled that result, with a single convolutional layer removed its performance degraded [24].

Since then, convolutional neural networks have become the bread and butter of recognition and classification tasks and have enabled real-time vision applications in smartphones, cameras, robots, and self-driving cars [25] and deep learning has become the new *it* thing. In 2012, at Andrew Ng's request, Hinton started the first-ever massive online open course on machine learning, *Neural Networks for Machine Learning*,[11] that enabled interested audience from all walks of life to learn about the emerging technology. The industry quickly seized the opportunity and acquired the talent behind the pioneering achievements—Hinton has been working part-time for Google since 2013 and is still trying to discover the learning procedure employed by the brain.

8.5 New Approaches

Having transitioned, in Hinton's own words, from the lunatic fringe to the lunatic core[12] he has gone full circle and hitched his wagon to an idea he feels extremely strongly about but not many people ready to put much stock in[6] **capsules**. A capsule is a group of neurons whose activity vector represents the instantiation parameters of a specific type of entity, such as an object or an object part; the length of that activity vector is used to represent the probability that the entity exists while its orientation represents the instantiation parameters [34]. They differ from neural networks in the activation—capsules get activated in response to comparison of multiple incoming pose predictions while neural networks compare a single incoming activity vector

[11] At Hinton's request the course has since been removed from Coursera due to being out of date but it can still be accessed from his webpage https://www.cs.toronto.edu/~hinton/coursera_lectures.html or YouTube.

[12] *Meet the man Google hired to make AI a reality*. For full article see https://www.wired.com/2014/01/geoffrey-hinton-deep-learning/.

and a learned weight factor [17] and represent a response to everything that is wrong with neural networks, especially in speech and object recognition where despite great success they still do not work as well as the brain does, which could be due to having fewer levels of structure.

Loosely inspired by minicolumns in the human visual system, capsules seek to rectify the inefficiencies by building an explicit notion of an entity into the architecture itself, thus providing equivariance where changes in the viewpoint, the biggest source of variation in images, lead to corresponding changes in neural activities because they capture the underlying linear structure, making use of the natural linear manifold that efficiently deals with variations in position, orientation, scale, and lightning.[13]

Introduced in 2011 as a simple way of recognizing wholes by recognizing their parts, Hinton, Krizhevsky, and Sida Wang proposed transforming autoencoders into first-level capsules which would have logistic recognition units that compute the outputs sent to higher levels and the probability that the capsule's entity is present in the input, and generation units that compute the capsule's contribution to the transformed image. From pixel intensities, first-level capsules extract explicit pose parameters of recognized fragments of their visual entities, externally supplied transformation matrices that learn to encode intrinsic spatial relationships between the parts and the whole are applied to those fragments and agreement of the poses predicted by active lower level capsules is then used to activate higher level capsules and predict the instantiation parameters of larger, more complex visual entities [14]. Sara Sabour, Nicholas Frosst, and Hinton further refined the system introducing an iterative routing-by-agreement mechanism for connecting layers in the feedforward network—dynamic routing—which enables capsules in the lower levels to send their output to parent capsules in higher levels that compute a prediction vector which, if their scalar product is large, increases the coupling coefficient for the chosen parent thereby increasing the contribution of that capsule and prediction with the parent's output [34]. More effective than max-pooling, dynamic routing, which allows capsules to selectively attend only to some active capsules in the layer bellow, enabled capsules to outperform similarly sized CNNs on affNIST and multiMNIST successfully recognizing multiple overlapping objects in images [34].

Exploring the potential of the new approach, the trio came up with a new matrix capsule system in which each capsule contains a logistic unit that represents the presence of an entity and a 4×4 pose matrix that captures its pose regardless of the viewpoint allowing the system to recognize objects with different azimuths and elevations [17]. A novel iterative routing procedure, based on the EM algorithm, routes the output of children capsules to parent capsules in an adjacent layer so that each active capsule receives a cluster of similar pose votes [17]. With these improvements, capsules outperform CNNs on the smallNORB dataset and display more robustness to white box adversarial attacks [17]. An unsupervised version of

[13] *What's wrong with convolutional nets?* Brain and Cognitive Sciences—Fall Colloquium Series Recorded December 4, 2014. For full lecture see https://techtv.mit.edu/collections/bcs/videos/30698-what-s-wrong-with-convolutional-nets.

capsules—Stacked Capsule Autoencoder—consists of two stages and presents an updated version devised by Adam Kosiorek, Sabour, Teh, and Hinton which no longer needs externally supplied transformation matrices but uses the image as the only input and no longer needs iterative routing because objects now predict parts [23]. In the first stage, part capsules segment the input into parts and poses, and reconstruct each image pixel as a mixture of affine-transforming learned templates while in the second stage object capsules arrange discovered parts and poses into a smaller set of objects thereby discovering the underlying structure [23]. Capsules are a work in progress but their built-in biases and unsupervised way of learning inspired by our own could contribute to the prevailing sentiment that the (next) "*revolution will not be supervised*".[14]

8.6 Five-Year Fog

Unknowable unknowns make long-term predictions about the future a fruitless endeavor. When asked to make them, Hinton prefers to use an analogy with driving a car during a foggy night—the distance that would otherwise be clearly visible and easily navigable becomes almost opaque due to the exponential effects of the fog that absorbs a fraction of photons per unit of distance.[15] Technology follows the same exponential progress and there is simply no telling where AI will end up in a couple of decades. Whatever the potential risks associated with it may be and regardless of the variation of the doomsday scenario one prefers to entertain, Hinton believes the problem is not technology itself—biases in neural networks are easy to fix, people present a greater challenge—but social systems that are rigged to benefit the top 1% at the expense of everybody else and one way to thwart that is through regulation of the use of AI, especially in weaponization, elections, and surveillance [4].

Fixing the system is a better long-term option than demonizing the technology and hindering its progress especially now when deep learning has gone mainstream with troves of researchers flocking to it. Hinton sees universities where young graduate students willing to question the basics and freely pursue truly novel ideas while being well advised by experts with similar views as more likely incubators of future advances than the industry [4]. Having time to read just enough to develop intuitions, ample insight to notice the flaws in the current approach and perseverance to follow those intuitions despite external influences is the only way to further the progress of deep learning,[6] given that this is exactly what Hinton has done it is not a bad example to approximate.

Neural networks started off as an underdog quickly dismissed as a curiosity that would never work and relegated to a fringe interest of especially perseverant individuals with a contrarian streak from various disciplines united under the header

[14]Yann LeCun.

[15]The final lecture of his *Neural Networks for Machine Learning* Coursera course. For full video see https://www.youtube.com/watch?v=IXJhAL6FEj0.

of cognitive science. Geoffrey Hinton is one of those pioneers whose belief in the connectionist approach has never been shaken. Earlier in his career, in the midst of general resistance, he stated: *But sooner or later computational studies of learning in artificial neural networks will converge on the methods discovered by evolution. When that happens a lot of diverse empirical data about the brain will finally make sense, and many new applications of artificial neural networks will become feasible* [7]. Due to his efforts and numerous inventions and improvements that followed from the continuous exchange of ideas between those he considers mentors and those he mentored, what he firmly believed in while many thought it was impossible is starting to happen sooner rather than later. Whatever emerges out of the fog will undoubtedly be influenced and shaped by his numerous and continued contributions.

References

1. Ackley DH, Hinton GE, Sejnowski TJ (1985) A learning algorithm for Boltzmann machines. Cogn Sci 9:147–169
2. Dahl GE, Sainath TN, Hinton GE (2013) Improving deep neural networks for LVCSR using rectified linear units and dropout. In: IEEE international conference on acoustic speech and signal processing (ICASSP 2013), pp 1–5
3. Dayan P, Hinton GE, Neal R, Zemel RS (1995) The Helmholtz machine. Neural Comput 7:1022–1037
4. Ford M (2018) Architects of intelligence: the truth about AI from the people building it. Packt Publishing, Birmingham, UK
5. Goodfellow I, Bengio Y, Courvile A (2016) Deep learning. The MIT Press, Cambridge, MA
6. Hinton GE (1981) Implementing semantic networks in parallel hardware. In: Hinton GE, Anderson JA (eds) Parallel models of associative memory. Lawrence Erlbaum Associates, pp 191–217
7. Hinton GE (1992) How neural networks learn from experience. Sci Am 267(3):145–151
8. Hinton GE (1999) Products of experts. In: Proceedings of the ninth international conference on artificial neural networks (ICANN 99), vol 1, pp 1–6
9. Hinton GE (2002) Training products of experts by minimizing contrastive divergence. Neural Comput 14:1771–1800
10. Hinton GE (2007) Learning multiple layers of representation. Trends Cogn Sci 11(10):1527–1554
11. Hinton GE (2007) To recognize shapes, first learn to generate images. In: Drew T, Cisek P, Kalaska J (eds) Computational neuroscience: theoretical insights into brain function. Elsevier, pp 535–548
12. Hinton GE (2014) Where do features come from? Cogn Sci 38(6):1078–1101
13. Hinton GE, Dayan P, Frey BJ, Neal R (1995) The wake-sleep algorithm for unsupervised neural networks. Science 268:1158–1161
14. Hinton GE, Krizhevsky A, Wand SD (2011) Transforming auto-encoders. In: International conference on artificial neural networks systems (ICANN-11), pp 1–8
15. Hinton GE, McClelland JL, Rumelhart DE (1986) Distributed representations. In: Rumelhart DE, McClelland JL (eds) Parallel distributed processing: explorations in the microstructure of cognition. Volume 1: foundations. MIT Press, pp 77–109
16. Hinton GE, Osindero S, Teh YW (2006) A fast learning algorithm for deep belief nets. Neural Comput 18:1527–1554
17. Hinton GE, Sabour S, Frosst N (2018) Matrix capsules with EM routing. In: International conference of learning representations (ICLR 2018), pp 1–15

18. Hinton GE, Salakhutdinov RR (2006) Reducing the dimensionality of data with neural networks. Science 313:504–507
19. Hinton GE, Sejnowski TJ (1983) Optimal perceptual inference. In: Proceedings of the IEEE conference on computer vision and pattern recognition, pp 448–453
20. Hinton GE, Srivastava N, Krizhevsky A, Sutskever I, Salakhutdinov RR (2012) Improving neural networks by preventing co-adaptation of feature detectors. arXiv:1207.0580, pp 1–18
21. Hinton GE, Zemel RS (1994) Autoencoders, minimum description length, and Helmholtz free energy. In: Cownan JD, Tesauro G, Alspector G (eds) Advances in neural information processing systems, vol 6. Morgan Kaufman, pp 1–9
22. Jordan MI, Ghahramani Z, Jaakkola TS, Saul LK (1998) An introduction to variational methods for graphical models. In: Jordan MI (ed) Learning in graphical models. Springer Netherlands, pp 105–161
23. Kosiorek AR, Sabour S, The YW, Hinton GE (2019) Stacked capsule autoencoders. arXiv:1906.06818 [stat.ML], pp 1–13
24. Krizhevsky A, Sutskever I, Hinton GE (2012) ImageNet classification with deep convolutional neural networks. In: Advances in neural information processing systems 25 (NIPS 2012), pp 1–9
25. LeCun Y, Bengio Y, Hinton G (2015) Deep learning. Nature 521:436–444
26. Longuet-Higgins HC (1973) Comments on the Lighthill report and the Sutherland reply. In: Artificial intelligence: a paper symposium. Science Research Council, pp 35–37
27. Mohamed A, Dahl GE, Hinton GE (2009) Deep belief networks for phone recognition. In: NIPS 22 workshop on deep learning for speech recognition, pp 1–9
28. Munakata T (2008) Fundamentals of the new artificial intelligence: neural, evolutionary, fuzzy and more. Springer, London
29. Nair V, Hinton GE (2010) Rectified linear units improve restricted Boltzmann machines. In: Proceedings of the 27th international conference on machine learning, pp 1–8
30. Neal RM, Hinton GE (1998) A view of the EM algorithm that justifies incremental, sparse and other variants. In: Jordan MI (ed) Learning in graphical models. Springer Netherlands, pp 355–368
31. Rosenfield E, Hinton GE (2000) In: Anderson JA, Rosenfield E (eds) Talking nets: an oral history of neural networks. The MIT Press, pp 361–386
32. Rumelhart DE, Hinton GE, Williams RJ (1986) Learning representations by back-propagating errors. Nature 323:533–536
33. Rumelhart DE, Hinton GE, Williams RJ (1986) Learning internal representations by error propagation. In: Rumelhart DE, McClelland JL (eds) Parallel distributed processing: explorations in the microstructure of cognition. Volume 1: foundations. MIT Press, pp 318–362
34. Sabour S, Frosst N, Hinton GE (2017) Dynamic routing between capsules. In: Conference on neural information processing systems (NIPS 2017), pp 1–11
35. Salakhutdinov R, Mnih A, Hinton GE (2007) Restricted Boltzmann machines for collaborative filtering. In: International conference on machine learning, Corvallis, Oregon, pp 1–8
36. Sejnowski TJ (2018) The deep learning revolution. The MIT Press, Cambridge, MA
37. Srivastava N, Hinton GE, Krizhevsky A, Sutskever I, Salakhutdinov R (2014) Dropout: a simple way to prevent neural networks from overfitting. J Mach Learn Res 15:1929–1958
38. Waibel A, Hanazawa T, Hinton GE, Shikano K, Lang KJ (1989) Phoneme recognition using time-delay neural networks. IEEE Trans Acoust Speech Signal Process 37(3):147–169

Chapter 9
Machine Learning and the Philosophical Problems of Induction

Davor Lauc

Abstract This chapter provides an analysis of the relationship of the traditional problems of justifying inductive inferences to the field of machine learning. After the summary of the philosophical problems of induction, text focus on the two philosophical problems relevant to the supervised and unsupervised machine learning. The former is a famous new riddle of induction devised by N. Goodman. The author argues that remarkable results in the theory of machine learning, no-free-lunch theorems are a formalisation of this traditional philosophical problem. Consequently, lengthy philosophical discussions on this problem are relevant to these results and vice versa. The later problem is the problem of similarity, as identified by N. Goodman and W. V. Quine. It is claimed that those discussions can help practitioners of unsupervised learning to be aware of its limitations.

Keywords Machine learning · New riddle of induction · No-free-lunch theorems · Problem of similarity · Philosophy of computing

9.1 Introduction

Human knowledge is tightly tied with inference and, arguably, all knowledge is based on some kind of inference. Consequently, an artificial system that poses knowledge or intelligence is also based on some mode of inference. Traditionally, all inference is partitioned into deduction and induction.

The former, deduction, has a special status. In well-formed deductive reasoning, the conclusion necessarily follows from the premises. A deduction can be seen as a truth-preserving machine, provide the true statement and it will yield the truth. The deductive inference is also very familiar to the computer science community, not only through mathematics and deductive systems but also through correspondence between programs and proofs, the famous Curry–Howard isomorphism.

D. Lauc (✉)
Faculty of Humanities and Social Sciences, University of Zagreb, Zagreb, Croatia
e-mail: dlauc@ffzg.hr

TerraLogix Group, London, UK

S. Skansi (ed.), *Guide to Deep Learning Basics*,
https://doi.org/10.1007/978-3-030-37591-1_9

However, the price that we pay for this special status of deduction, of its infallibility, is enormous. Although some deductive inferences such as mathematical proof took years or even centuries to be completed, the deduction does not yield any new information in an absolute sense. As it can be easily demonstrated within the framework of semantic information theory [6], the conclusion of a deductively valid argument never contains more information than the premises. In the philosophical jargon, this result is usually called the *scandal of deduction*.

Thus, to get new knowledge in an absolute sense, we need the following form, the non-deductive forms of inference. In the modern sense, all of the non-deductive reasoning is considered as a form of inductive reasoning. The standard definition, complementary to the definition of deduction, is that inductive inference is the one where it is possible to get a false conclusion from the true premises. Naturally, this definition covers a wide range of inferences.

In the philosophical context, it is useful to distinguish two groups of the problem related to inductive inference. The first group is related to developing methods of induction or at least method of distinguishing better from worse inductive inferences. Although there is still a debate in the philosophy of developing such *inductive logic*, it seems that the leading role in this area has been overtaken by other disciplines. The science of statistics, and even machine learning, can be considered as an endeavour to develop such inductive inference.

The other group of problems belongs to the more hardcore philosophical issue, the challenge of *justifying* the inductive inference. To some extent, statistics tackle the problem of justification, and fields like *formal learning theory* are concerned with such justification. However, those problems are still mainly in the realm of philosophy. It is beneficial to contrast those two groups using the recent theory of human reasoning developed by psychologists Hugo Mercier and Dan Sperber, which is depicted in Fig. 9.1 (Mercier and Sperber, 2017).

If their theory is empirically adequate, the reasoning is deeply embedded in several categories of inference, and the justification of reason is just one part of it. Even if this theory is not an adequate description of the way humans infer and reason, in this context, it is a restatement of well-known Reichenbach distinction between the *context of discovery* and the *context of justification*. We will ignore how a human or an artificial reasoner comes to inductive inferences. Our focus will be on the philosophical problem of how we can justify such conclusions, and if it is possible at all.

9.2 What Are the Philosophical Problems of Induction?

As the problem of induction is one of the most significant philosophical issues in the last few hundred years, consequently, tons of literature about it has been created. So, it is entirely beyond the scope of this text to provide a comprehensive review of it. We will focus on a bird's eye review from the perspective of machine learning of the three most pressing problems of induction, which are related but distinct.

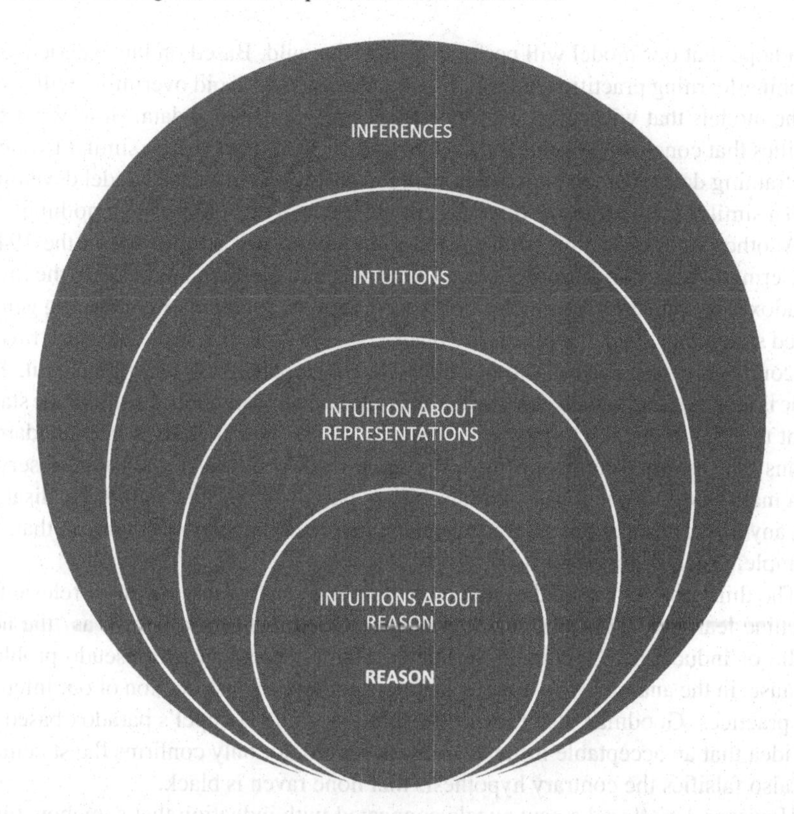

Fig. 9.1 Mercier and Sperber embedding of reason

The problem of induction, or the classical formulation of the problem, is also called the Hume's problem of induction. David Hume formulated it in the seventeenth century, and in modern terminology it can be shortly reformulated as follows. He poses the question of how we can justify inductive inference, how we can conclude from the observed to the unobserved. There are two ways, demonstrative and probable in Hume's terminology, that is deductive or inductive. If we take the first way, we would need to presuppose as a premise that unobserved will be similar to observed, but this is what we are trying to prove precisely. If we take the second way, we are using the principles of reasoning we are trying to justify. So, we are using unacceptable circular reasoning. As that exhaust all possibilities of justifying the inductive inference, we must conclude that it is unjustifiable. It is vital to notice that Hume does not deny prevalence or importance of such reasoning; he is only stating its unjustifiability.

In the context of machine learning, we can understand this classical problem as a generalisation of the bias-variance trade-off. We can use high regularisation of the model to keep it simple (other unjustified assumption), set hyperparameters and test it on the held-out part of the dataset, use cross-validation and other tricks of trade and then hope that our model will perform well in the wild. Based on the experience of machine-learning practitioners, using such techniques to avoid overfitting will result in the models that will perform better on new, yet unobserved data. However, what justifies that conclusion? A natural law that the real-life data will be similar to one in our training dataset or the experience with the past performance of model developed with a similar methodology? If we accept the Hume's dichotomy, then nothing.

Another well-known problem related to induction was formulated in the 1940s by German-American philosopher Carl Gustav Hempel, often labelled as "the raven paradox". In a nutshell, Hempel is concerned with the problem of confirming generalised statements, like "all ravens are black". It is natural to accept that black ravens are confirming such generalisation while raven of any other colour will falsify it. The issue is that, logically, such generalisation is equivalent to its contraposition, the statement that "all nonblack things are nonravens". This logical equivalence standardly means that the two statements have the same truth conditions; the same observed data makes both of them true and false. The "paradox" unfolds with the realisation that any black thing or any nonravens make the previous statement right, so that, for example, green apple confirms it.

The third problem related to inductive inference is arguably the most relevant to machine learning. It is formulated by Nelson Goodman and is known as "the new riddle of induction". Goodman considered Hume's problem as a pseudo-problem because, in the analogy with deduction, there cannot be a justification of our inferential practices. Goodman also provided a solution to the Hempel's paradox based on the idea that an acceptable instance (a black raven) not only confirms the statement but also falsifies the contrary hypothesis that none raven is black.

However, he offered a new puzzle connected with induction that somehow turns Hume's problems on its head. In the sense, that it is not the case that none of the inductive inferences are justified, but that it is too many of them. His famous example is that, based on many observed green emeralds, and none emeralds of other colour, we can naturally conclude that all emeralds are green. However, this is not the only conclusion we could reach. Let us define a new property called "grue". The property is true of all things that are blue and observed before some point in future, for example before the year 2100, or observed after this time point and blue. Based on the same data, we could conclude that all emeralds are "grue", and of course numerous other statements generated in this fashion. Goodman challenge is to distinguish properties we could use in inductive inference like green, from the one we cannot. He named the former projectable and the latter non-projectable. In spite of numerous discussions on this topic in the last 50 years there is still little agreement about how to solve his problem.

It is worthwhile to analyse a similar argument, without the fancy neologism, made by Goodman earlier:

> Suppose we had drawn a marble from a certain bowl on each of the 99 days up to and including VE day, and each marble drawn was red. We would expect that the marble drawn on the following day would also be red. So far all is well. Our evidence may be expressed by the conjunction "$R_a 1$ & $R_a 2$ & & & $R_a 99$," which well confirms the prediction "$R_a 100$" But increase of credibility, projection, "confirmation" in any intuitive sense, does not occur

in the case of every predicate under similar circumstances. Let "S" be the predicate "is drawn by VE day and is red, or is drawn later and is non-red." The evidence of the same drawings above assumed may be expressed by the conjunction "$S_a 1 \& S_a 2 \&\&\& S_a 99$." By the theories of confirmation in question, this well confirms the prediction "$S_a 100$" but actually we do not expect that the hundredth marble will be non-red. "$S_a 100$" gains no whit of credibility from the evidence offered hundredth marble will be non-red.

From this quotation, the relevance to machine learning is more evident. By using different learning setups and algorithms we can reach different models that are indistinguishable by our dataset. We will analyse this into more details later in the text.

9.3 Why Are the Philosophical Problems of Induction Relevant to Machine Learning?

Leslie Gabriel Valiant, a British computer scientist, claims that "...induction is a pervasive phenomenon exhibited by children. It is as routine and reproducible a phenomenon as objects falling under gravity." [27]. To extend his metaphor further, although effects of gravitation were of course known in antiquity, we were not satisfied by the Aristotelian explanation that a body falls to the earth because it is their natural place, for this is where it belongs. Neither we were completely satisfied by the pretty accurate Newtonian description of how gravity works, nor with better description and understanding made by general relativity. We still want to have enhanced understanding of the nature of a phenomenon of gravity, not only to account for anomalies and discrepancies like extra energetic photons and provide better description and prediction but for the sake of understanding it as such.

By analogy, as machine learning is the venture of creating computer systems that learn from data, such system can be viewed as an inductive machine, as a device that is based on and performs inductive inferences. We are not, and should not be, satisfied by the fact that the systems we are developing are getting better at their tasks, that perform inductive inference well. We would like to know more, to understand the whys of induction and learning, to better understand the limits and explainability of it, not only of the present learning systems but any future systems as well. As we are putting more trust in machine learning systems and integrating them into our lives, from health to legal system, it is crucial to understand their foundation and limits more thoroughly.

It may happen, as it already happened many times in the development of philosophy and sciences, from physics to psychology, that the problems of induction detach from the mothership of philosophy. It may become a problem of some particular science, or even new science of it own, like the formal learning theory. However, relevant questions about induction are still so open-ended, and there is still so little understanding of the exact nature of those questions, methods to approach them and possible solution. So, it is expected that the problem will stay "philosophical" for at least some time.

So even if philosophy and philosophical problems are at present, less relevant to the actual practice of science and technology than they used to be, it is plausible to argue that the minimal relevance of philosophical problems is to avoid rediscoveries and repeat mistakes made by philosophers. Namely, in the two-and-a-half millenniums of philosophical endeavours, many questions were raised, many problems and several solutions were discussed, and numerous pro-and-contra arguments were examined. Almost by definition, many of the discussed problems are pseudo-problems, many of the solutions are bogus, and many of the arguments are unconvincing. However, by the same token, many of the arguments are solid and show that some theories are implausible to be valid, so the least we can learn from philosophy is not to repeat same mistakes again and again.

Let us exemplify this with the claim related to the problem of induction and machine learning made 10 years ago by C. Anderson, former editor-in-chief of Wired Magazine. In his influential and provocative article, inspired by big-data hype, "The End of Theory: The Data Deluge Makes the Scientific Method Obsolete", he states that fundamental scientific methodology—hypothesise, model, test—is now obsolete due to a vast amount of data. He claims that "with enough data, the numbers speak for themselves", and that petabytes of data allow us to say that correlation is enough, that correlation *is* causation.

For philosophers, this is not a new claim. It is just reiteration of one of the many arguments made by empiricist and rationalists in 25 old debate on the source and justification of human knowledge. Thesis formulated by Anderson was most notably made in seventeenth century by Francis Bacon in his *Novum Organum*. Bacon's main point was that contrary to the then-dominant Aristotelian scientific method based on deduction. He argued that scientific knowledge should be based on experimental data. In his famous metaphor, scientists are bees that unlike spiders the rationalist or ants the pure empiricist are taking the middle course. They take materials from the flowers (observations) and "transforms and digests it by a power of its own" [2]. In the contemporary analogy, we only need to provide enough data to our deep learning models, and they will flourish and produce knowledge.

Among many arguments against this position, let us consider the one stated by Sir Karl Poppers. He can be considered as, at least philosophical, father of the scientific method attacked by Anderson.

> … the belief that we can start with pure observations alone, without anything in the nature of a theory, is absurd; as may be illustrated by the story of the man who dedicated his life to natural science, wrote down everything he could observe, and bequeathed his priceless collection of observations to the Royal Society to be used as inductive evidence. This story should show us that though beetles may profitably be collected, observations may not. Twenty-five years ago I tried to bring home the same point to a group of physics students in Vienna by beginning a lecture with the following instructions: 'Take pencil and paper; carefully observe, and write down what you have observed! They asked, of course, what I wanted them to observe. Clearly, the instruction, 'Observe!' is absurd. … Observation is always selective. It needs a chosen object, a definite task, an interest, a point of view, a problem. Moreover, its description presupposes a descriptive language, with property words; it presupposes similarity and classification, which in its turn presupposes interests, points of view, and problems. … objects can be classified, and can become similar or dissimilar, only in this

> way – by being related to needs and interests. ... a point of view is provided ... for the scientist by his theoretical interests, the special problem under investigation, his conjectures and anticipations, and the theories which he accepts as a kind of background: his frame of reference, his 'horizon of expectations'. [22]

Digested reiteration of this Popper's argument can be heard in the machine learning community under the dictum *no learning without bias* [5, 15].

Among many philosophical arguments related to machine learning, in this work, we will focus on the two examples. The first one is related to the supervised learning and our thesis that the well-known no-free-lunch theorems are just reiteration of famous Nelson Goodman's New Riddle of Induction problem. The latter is related to unsupervised learning and the relevance of the Willard Van Orman Quine analysis of the problem of similarity to the problem of clustering.

9.4 Supervised Learning and the New Riddle of Induction

It is a pearl of received wisdom the no-free-lunch (NFL) theorems—the great negative results in machine learning—as a reiteration or even a formalisation of the Hume's problem of induction.[1] The distinguished machine-learning researchers like Christophe Giraud-Carrier and Pedro Domingos state, respectively:

> It then becomes apparent that the NFL theorem in essence simply restates Hume's famous conclusion about induction having no rational basis... [8]

also,

> ...This observation was first made (in somewhat different form) by the philosopher David Hume over 200 years ago, but even today many mistakes in machine learning stem from failing to appreciate it." [4]

Even the originator of the first form of the NFL theorems, David Wolpert, 15 years after he proved the theorem, joins the information cascade and claims that:

> ...these original theorems can be viewed as a formalisation and elaboration of concerns about the legitimacy of inductive inference, concerns that date back to David Hume... [30]

This chapter argues that the NFL theorems, although vaguely connected to the classical philosophical problem of induction, do not restate the Hume's problem, but rather the associated Nelson Goodman's argument.[2] We claim that NFL theorems are closely related to Goodman's new riddle of induction (NRI), to the extent that they are one possible formalisation of the riddle. Additionally, we would like to pose the question of the relevance of the NFL theorems to the lengthy philosophical discussion on NRI, as the relationship is yet to be researched. The related, reversed,

---

[1]This part of the text is an extended version of the author's text "How Gruesome are the No-free-lunch Theorems for Machine Learning?" [16].

[2]We suppose that the new riddle is a different issue from the classical problem of induction, what is the received position with a few notable exception.

the issue is the relevance of NRI to NFL and the question as to whether the machine-learning community could benefit from the almost 70 years of fruitful discussion about Goodman's argument.

9.4.1 No-Free-Lunch Theorems

The first form of NFL theorem was proven by Wolpert and Macready, 1992, in the context of computational complexity and optimisation research [28, 31]. He later proved the variant of the theorem for the supervised machine learning [29]. For the sake of our argument, we will sketch the proof of the simplified version of the theorem for supervised learning based on the work of Cullen Schaffer [25].

In the simplest, discrete settings of the machine learning of a Boolean function, training data X consists of the set of binary vectors representing a set of attributes that are true or false for each instance of binary function—concept. Each vector is labelled as a positive or negative example of a concept we want to learn. The machine-learning algorithm L tries to learn a target binary function y; that is, it tries to learn a real concept from this set of examples. The training dataset is always finite with some length n, and the relative frequency of data feed to L is defined by probability distribution D. In a context more familiar to the philosophers, this problem of machine learning can be seen as a guessing a true form of a large n-ary truth function from the partial truth-table, where most of the rows are not visible.

The key performance indicator of a machine learner is a generalisation performance, with the accuracy of the learner found within the data outside the training dataset. Modern machine-learning algorithms can easily "memorise" data from the training dataset, and perform poorly on the "unseen" data, leading to the problem known as overfitting. So, the success of the learner is measured by how well it will generalise, and how well it performs on the novel data. In the simple setting of binary concept learning, the baseline of the generalisation accuracy of a learner, $GP\ (L)$ is at the level of a random guess, with the accuracy of the novel data being 0.5. Such performance is the result we will expect on average if we use the toss of a coin to decide, for an unseen example, whether it belongs to our target concept or not. Obviously, we want any learner to perform better than this.

The NFL theorem claims that, for any learner L, given any distribution D and any $n\ of\ X$

$$\frac{\sum_{f \in Y} GP(L)}{|Y|} = 0.5,$$

where Y is a set of all target functions, all possible concepts that can be learned.

So, the theorem states that, on average, the generalisation performance of any learner is no better than random guessing. All learning methods, from the simple decision trees to the state-of-the-art deep neural networks, will perform equally when all possible concepts are considered.

This result, unanticipated at least on the first sights, does bear some resemblance to the discrepancy between the results of the argument and our expectations in the case of the Hume's argument. However, it does not claim that we cannot learn anything from the training data or experience, but that we can learn everything, which is, arguably, the point of Goodman's argument. The resemblance to the NRI will be more evident from the sketch of the proof of the NFL theorem. The basic idea is straightforward: for any concept that the learner gets right, there is a concept that it gets wrong or, in Goodman's lingo, for every "green" concept there is a "grue" concept. The "grue" concept is constructed similar to the NRI argument, in that it agrees on all observed data—data in the training dataset—with the "green" concept, and it is bent on all non-observed data.

More formally, for every concept C that L learns to classify well—say it classifies m novel examples accurately—there is a concept C' that L learns where all m examples will be misclassified. C' is constructed as follows:

$$C' = \begin{cases} C & \text{if } x \in X \\ \neg C & \text{if } x \notin X \end{cases}$$

Visually, this simple construction of the concept C' corresponds to the Wittgenstein–Goodman "bent predicate" [1], where X represents observed data (training dataset) and X′ unobserved data.

From the perspective of the primary measure of the success of the learning—generalisation accuracy, for every accuracy improvement a over the baseline for a concept C, there is a concept C' that will offset the improvement of the accuracy by $-a$. Consequently, the improvement in accuracy for any learner over all possible concepts is zero. It is possible to generalise this result to the more general learning settings, and many extensions of the theorem are proven [13, 14] (Fig. 9.2).

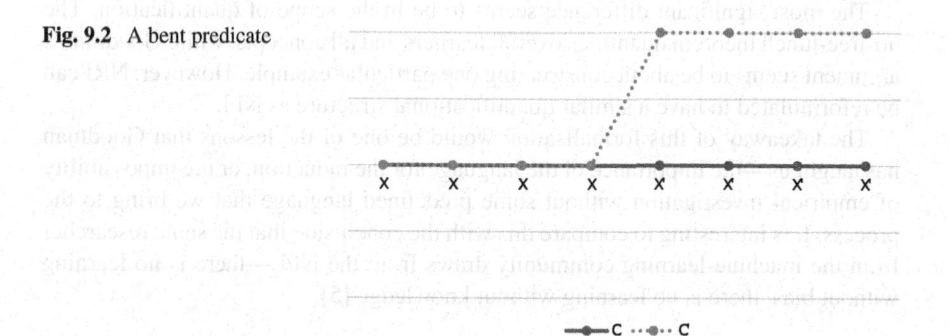

Fig. 9.2 A bent predicate

9.4.2 *Is the No-Free-Lunch Theorem the New Riddle of Induction?*

Although NFL bears a strong resemblance to NRI, it seems worthy to analyse differences and similarities between these two results. Let us start with similarities. Both arguments are about inductive inference, about inferring from the known to the unknown, and from the observed to the unobserved data. Both arguments imply that there are too many inductive inferences that can be inferred. Furthermore, both arguments seem to draw empirically inadequate conclusions, contrary to scientific practice and reasonable expectation. Nobody is expected to conclude that all emeralds are *grue*, and neither that random guess is an inductive strategy as good as any other.

The fundamental resemblance is in the construction of the arguments, the split of the evidence and the bend in the unobserved data. In most of the NRI arguments, we split the evidence into observed and unobserved (sometimes to some point in the future). Equally, in the NFL, data is split into observed, training dataset and the unobserved data to which the learning algorithm should generalise. In both arguments, the other counter-concept, *grue* or C', is constructed in the same manner. It agrees on the observed data and bends on the unobserved data.

Regarding the differences, firstly, there is a difference in the argument contexts. NRI was made in the philosophical, theoretical context of the logic of confirmation and pragmatic vindication of induction, while the NFL was made in the technological context of artificial intelligence and computing. The aim of the arguments also differs, at least at first glance. The intention of NRI, at least in Goodman's original form [9, 11], was to recognise one of the problems in the logic of confirmation—the demarcation between projectable and non-projectable predicates. On the other hand, the objective of the NFL was to demonstrate that there is no single best algorithm, initially for the optimisation and search, and later for supervised learning.

The most significant difference seems to be in the scope of quantification. The no-free-lunch theorem quantifies overall learners and all concepts, while Goodman's argument seems to be about constructing one particular example. However, NRI can be reformulated to have a similar quantificational structure as NFL.

The takeaway of this formalisation would be one of the lessons that Goodman has taught us—the importance of the language for the induction, or the impossibility of empirical investigation without some predefined language that we bring to the process. It is interesting to compare this with the conclusion that the same researcher from the machine-learning community draws from the NFL—there is no learning without bias, there is no learning without knowledge [5].

9.5 Unsupervised Learning and the Problem of Similarity

Willard Van Orman Quine, one of the greatest philosophers of the twentieth century, accepted Hume's challenge to inductive inference and agreed that justification could be provided neither by experience nor a priory—"The Humean predicament is the human predicament" [21]. Quine is trying to provide, in the framework of his naturalistic epistemology, an explanation of our inductive practices based on evolution. He claims that all learning is based on conditioning, thus on induction, and therefore induction itself cannot be learned:

> … the instinct of induction: the instinct to expect perceptually similar stimulations to have similar sequels. … Philosophers have marvelled that expectation by induction, though fallible, is so much more successful than random guessing. This is explained by natural selection …" [20]

Regarding the other two problems, Quine notices that the Hempel's problem, the raven paradox, is reducible to Goodman's problem [23]. Precisely, a complement of a projectible predicate (black–nonblack), does not need to be, also almost never is, projectible. His simple solution to the new riddle of induction is that "green emeralds" are more *similar* to each other than the "grue emeralds". However, this reduction of the problems of induction to the problem of similarity is not an easy solution, because "..the dubious scientific standing of a general notion of similarity … the dubiousness of this notion is itself a remarkable fact" [23].

Quine was not the first philosopher to tackle the concept of the similarity, among others Leibniz, Hume and Goodman were analysing it. Leibniz considered similarity as a weakened version of his famous principle of identity of indiscernibles. Two objects (substances in his terminology) are identical if they share all properties, two objects are similar if they share at least one [18].[3] Likewise, Hume position of similarity (resemblance) is that degree of similarity of two concepts depends on the number of "simple ideas", properties, that they share [12], the approach that is today called *the common attribute view of similarity*.[4]

In the context of machine learning problem of similarity emerges primarily in unsupervised learning, although it is relevant to supervised learning tasks as well. In clustering problem, the primary technique of unsupervised learning, similarity (closeness, distance) metric is input to clustering algorithms as well as to all its performance metrics. There are several philosophical and practical problems with supervised learning even once the distance metrics are defined, reaching from determining number of clusters to the interpretation of clusters, but here we will focus on the concept of similarity itself as a prerequisite of any unsupervised learning technique.

Let us illustrate the problem of similarity with a simple, practical application, that is the problem of reasoning with dates, the temporal references with the granularity of

[3]Rodriguez-Pereyra claims there are strong philosophical grounds that Leibniz thought that the similarity of substances does not derive from the similarity of their properties (accidents) [24].

[4]Although there are readings of Hume that suggest that Hume's position on similarity is closer to Goodman's view [7].

days. Date similarity seems a trivial problem as we can use simple absolute difference in days between any two dates as similarity metrics. However, how to extend this the similarity metrics to the inexact temporal expressions, dates determined by inexact expressions such as “the beginning of the twentieth century” or “the date of birth of the person who died in 1875”? It is natural to represent such dates, that often occur in practical application, using discrete probability distribution over the range of relevant period. So, for example, we could represent exact dates with distribution where total probability mass is assigned to one day. If the date is from an unreliable source, we can assign only 0.1 mass to it and spread the rest to the neighbouring dates according to some probability distribution. In the same manner, we can represent expressions like “XVII century” or “beginning of the XVIII” century (see Fig. 9.3).

So, if we want to cluster such dates, what similarity metrics should we use? There is no simple distance in parameters of distribution because many different distributions that need to be used. There are many similarity measures and distance function between probability distribution in different scientific fields. In statistics and probability theory, there are distance correlation, Bhattacharyya distance, f-divergences like Kullback–Leibler, Kolmogorov–Smirnov and many others [3]. There are information-theoretical distance/similarity measures like mutual information or Jensen–Shannon divergence, as symmetric version of Kullback–Leibler divergence [19]. The principal problem with all those distances is that they do not satisfy the underlying intuitive semantics of the inexact date’s comparison.

Even worse, similarity metrics as perceived by the field experts, historians, in this case, do not even satisfy the essential requirement of any metrics—reflexivity. Professionals perceive narrow ranges of dates as more similar to itself than the larger ones, so, for example, the year 1941 is more similar to itself than the twentieth century [17]. In the context of cognitive sciences, Tversky studied extensively this problem of subjective and highly contextual perception of similarity [26]. He demonstrated that human judgement of similarity is not symmetrical, as test subjects perceived that North Korea is more similar to China than vice versa.

As it is often the case, this empirical research was preceded by philosophical analysis. Goodman wrote off the concept of similarity in both philosophically and scientifically use:

> ...Consider baggage at an airport checking station. The spectator may notice shape, size, colour, material, and even make of luggage; the pilot is more concerned with weight, and the passenger with destination and ownership. Which pieces are more alike than others depends not only upon what properties they share, but upon who makes the comparison, and when. [10]

The takeaway from this and related discussions for the unsupervised learning could be expressed in Quinean lingo like: there is no “fact of the matter”, no real ground truth for the similarity metrics, hence for any association or clustering learning. Those structures are not something objective in the nature that science can discover as we invented it, and we should be aware of contextuality and limitation of it.

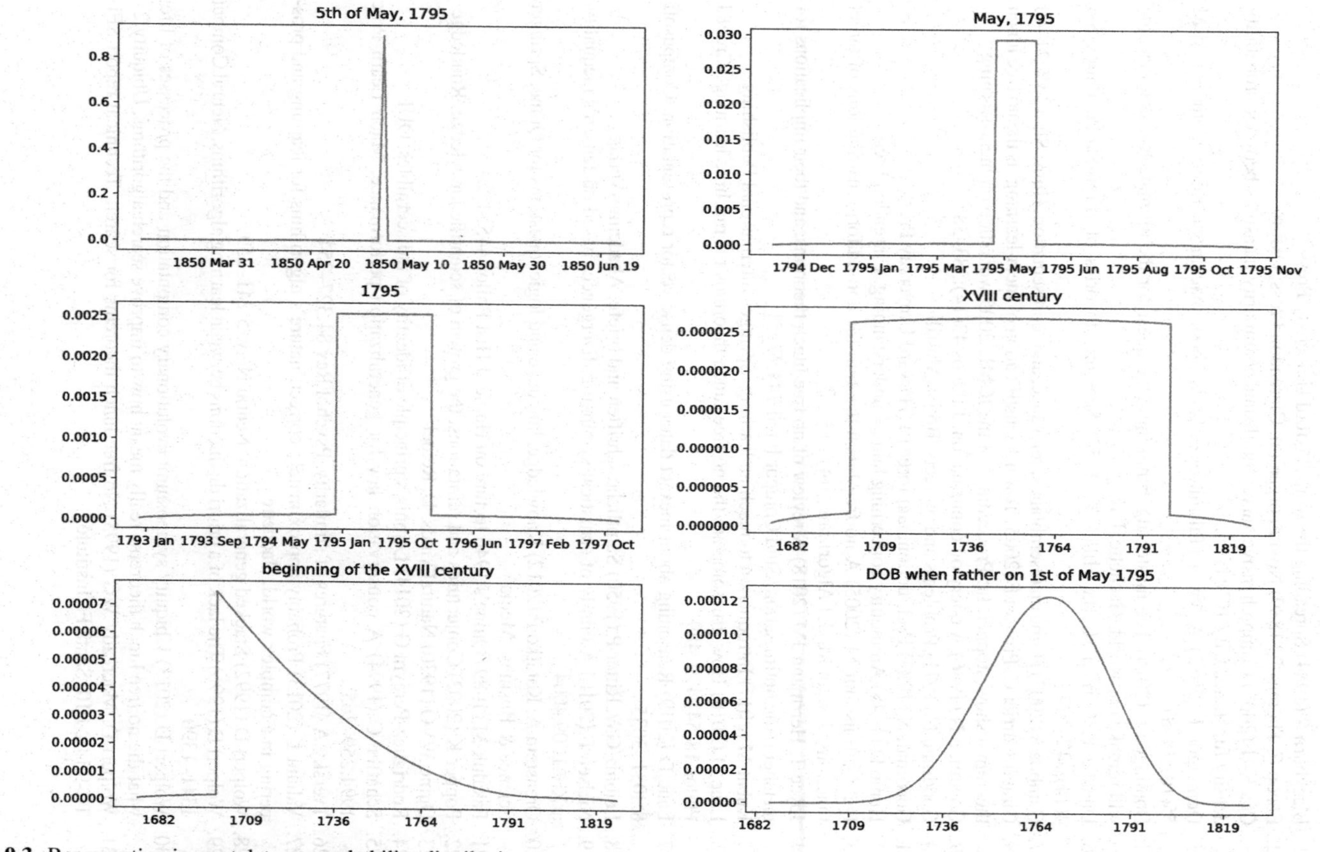

Fig. 9.3 Representing inexact dates as probability distributions

References

1. Blackburn S (1984) Spreading the word. Oxford University Press
2. Bacon F, Thomas F (1889) Novum organum. Clarendon Press, Oxford
3. Cha S-H (2007) Comprehensive survey on distance/similarity measures between probability density functions. City 1(2):1
4. Domingos P (2012) A few useful things to know about machine learning. Commun ACM 55(10):78–87
5. Domingos P (2015) The master algorithm: how the quest for the ultimate learning machine will remake our world. Hachette, UK
6. Dretske F, Carnap R, Bar-Hillel Y (1953) Semantic theories of information. Philos Sci 4(14):147–157
7. Gamboa S (2007) Hume on resemblance, relevance, and representation. Hume Stud 33/1:21–40
8. Giraud-Carrier C, Provost F (2005) Toward a justification of meta-learning: is the no free lunch theorem a show-stopper. In: Proceedings of the ICML-2005 workshop on meta-learning
9. Goodman N (1946) A query on confirmation. J Philos 43(14):383–385
10. Goodman N (1971) Problems and projects. Bobbs-Merrill
11. Goodman N (1983) Fact, fiction, and forecast. Harvard University Press
12. Hume D (1748) An inquiry concerning human understanding. Clarendon Press
13. Igel C, Toussaint M (2005) A no-free-lunch theorem for non-uniform distributions of target functions. J Math Model Algorithms 3(4):313–322
14. Joyce T, Herrmann JM (2018) A review of no free lunch theorems, and their implications for metaheuristic optimisation. Stud Comput Intell 744:27–51
15. Kubat M (2017) Introduction to machine learning. Springer International Publishing
16. Lauc D (2018) How gruesome are the no-free-lunch theorems for machine learning? Croat J Philos 18(54):479–489
17. Lauc D (2019) Reasoning about inexact dates using dense vector representation. Compusoft 8:3031–3035
18. Leibniz GW, Ritter P (1950) Sämtliche schriften und briefe. Akademie-Verlag
19. Nielsen F (2010) A family of statistical symmetric divergences based on Jensen's inequality. arXiv:1009.4004
20. Orenstein A, Kotatko P (2012) Knowledge, language and logic: questions for Quine. Springer Science & Business Media
21. Pakaluk M (1989) Quine's 1946 lecture on Hume. J Hist Philos 445–459
22. Popper K (2002) Conjectures and refutations: the growth of scientific knowledge. Routledge
23. Quine WVO (1970) Natural kinds. D. Reidel
24. Rodriguez-Pereyra G (2014) Leibniz's principle of identity of indiscernibles. OUP
25. Schaffer C (1994) A conservation law for generalization performance. Mach Learn Proc 1994:259–265
26. Tversky A (1977) Features of similarity. Psychol Rev 84:327–354
27. Valiant L (2013) Probably approximately correct: nature's algorithms for learning and prospering in a complex world. Hachette
28. Wolpert D (1992) Stacked generalization. Neural Netw 5:241–259
29. Wolpert D (1996) The lack of a priori distinctions between learning algorithms. Neural Comput 1341–1390
30. Wolpert D (2013) Ubiquity symposium: evolutionary computation and the processes of life: what the no free lunch theorems really mean: how to improve search algorithms. Ubiquity 2
31. Wolpert D, Macready WG (1995) No free lunch theorems for search. Technical report SFI-TR-95-02-010, Santa Fe Institute

Chapter 10
The Artificial Intelligence Singularity: What It Is and What It Is Not

Borna Jalšenjak

Abstract The topic of this chapter is artificial intelligence singularity. The artificial intelligence discussed in this chapter is general artificial intelligence that might appear as the result of numerous self-improvement cycles which result in a highly advanced version of the intelligence that people created. Using insights from classical philosophical anthropology into what characteristics are present in live beings author suggests that there are enough similarities between beings which are traditionally considered to be alive and the possible advanced general intelligence to claim that through analogy we should prepare ourselves to accept this hypothetical artificial intelligence as a being which is alive.

Keywords Life · Artificial intelligence · Philosophical anthropology · Singularity · Responsibilities

10.1 Introduction

In the city of Zagreb (in Croatia) there was a saying *Run, people—the car is coming!* Urban legend says that people used to shout this in times when cars were a novelty to warn others of the imminent danger. Today it seems that it would be appropriate to shout *Run, people—AI is coming*! I think this illustrates two things. One, people do not change much; we were and will be vary, of new things. Also, sometimes we feel at unease, without a good reason, in the face of novelty. There is no need to be afraid of everything new. In this chapter, I will try to present and discuss some of challenges and questions regarding artificial intelligence (AI), intelligence explosion, singularity and the future as I see them from the perspective of philosophical anthropology (PA).

Few introductory remarks. First, in this chapter, discussions about AI are about the so-called general artificial intelligence. This is important because there can be an AI which is great at making hamburgers but is not good at anything else. Such AIs might be called domain-specific. Different from them, the focus of this chapter is an

B. Jalšenjak (✉)
Zagreb School of Economics and Management, Zagreb, Croatia
e-mail: bjalsenjak@gmail.com

S. Skansi (ed.), *Guide to Deep Learning Basics*,
https://doi.org/10.1007/978-3-030-37591-1_10

AI that is not subject-specific, or for the lack of a better word, it is domainless and as such it is capable of acting in any domain. Something like human intelligence. Second note, domainless AI is probably something that most people think of when discussing the risks or the benefits of an AI, but at the moment there is almost no actual work being done on such a version of an AI, and researchers are directing the majority of their efforts at domain-specific AIs [1, p. 2].

Nonetheless, as long as there is a chance that general AI might exist sometime in the future, as seems to be the case based on estimation polls done in the science community [2], the topic seems to be too important not to explore.

The discussion about the possibility and actual way that general AI might emerge is outside of the scope of this chapter. There are many volumes already dedicated to this topic.[1] This chapter tries to add to the discussion starting from a question that is asked in the [4, pp. 3–4]. In that paper, authors give a list of question and encourage authors from different fields to answer them. One of the questions that they ask is: "What are the necessary and sufficient conditions for machine intelligence to be considered to be on par with that of humans? What would it take for the "general educated opinion [to] have altered so much that one will be able to speak of machines thinking without expecting to be contradicted" (Turing 1950, p. 442)?" [4, pp. 3–4] My opinion as to what will move general educated opinion to be more ready to accept machines as thinking beings is to start looking at them as being alive.

10.2 Singularity and AI Singularity

There are diverse views what does singularity mean. Different understandings of the concept come from various sources spanning science fiction, futurology, technical fields, everyday speech, philosophy, and computer sciences. For example, when Alan Turing [5] discussed the concept of singularity, he wrote: "(…) for it seems probable that once the machine thinking method had started, it would not take long to outstrip our feeble powers. There would be no question of the machines dying, and they would be able to converse with each other to sharpen their wits. At some stage, therefore, we should have to expect the machines to take control (…)". Another author, who actually introduced the concept of singularity, Vernor Vinge [6] writes: "We will soon create intelligences greater than our own. When this happens, human history will have reached a kind of singularity, an intellectual transition as impenetrable as the knotted space-time at the center of a black hole, and the world will pass far beyond our understanding." In the general public probably best-known author on singularity is the inventor Ray Kurzweil. He defines singularity as [7, p. 7]: "It's a future period during which the pace of technological change will be so rapid, its impact so deep, that human life will be irreversibly transformed. Although neither utopian nor dystopian, this epoch will transform the concepts that we rely on to give meaning to our lives, from our business models to the cycle of human life,

[1] As an example please see [3].

including death itself. Understanding the Singularity will alter our perspective on the significance of our past and the ramifications for our future. To truly understand it inherently changes one's view of life in general and one's own particular life. I regard someone who understands the Singularity and who has reflected on its implications for his or her own life as a 'singularitarian'."

From these examples it can be seen that the word singularity is used differently by different people. Today it has even become a symbol of a sort of a popular movement championed by Kurzweil and distancing itself from the academia [1].

In this chapter, we will understand singularity as an AI singularity and following what Chalmers [8, p. 7] says what might happen when machines become more intelligent than humans: "(…) this event will be followed by an explosion to ever-greater levels of intelligence, as each generation of machines creates more intelligent machines in turn." Chalmers, in the cited article, calls this ever improving machine and the resulting intelligence explosion the singularity. Said in a more succinct way, once there is an AI which is at the level of human beings and that AI can create a slightly more intelligent AI, and then that one can create an even more intelligent AI, and then the next one creates even more intelligent one and it continues like that until there is an AI which is remarkably more advanced than what humans can achieve. In this chapter we will use the concept singularity in just explained way.

10.3 Various Questions and Philosophical Anthropology

Leaving aside the presumptions and the prognosis how fast is singularity achievable and what are the reasons why it might or might not happen, there are many interesting philosophical questions entangled with the notion of singularity. As I see them, they can be grouped into several categories. The first group of questions is part of philosophical anthropology and analogous to it to computer anthropology (for the lack of better word and suffering from heavy anthropomorphization).[2] Second group of questions will belong to the category of ethics. And the third group of questions will belong to the field of motivation and action theory. In this chapter, the focus is on the first group of questions, while others are only briefly mentioned.

Generally saying, PA is one of the classical disciplines in philosophy. Following the usual categorization of sciences using their material and formal object, PA is concerned with all properties of human beings as its material object and its formal object is their existence. The emphasis is on the most significant properties of human beings.

One traditional way of discussing and teaching topics belonging to PA was using a list of theses, i.e., statements about a particular state of things, concerning different

[2] A note is needed here to say that there is no reason to think that advanced AI will have the same structure as human intelligence if it even ever happens, but since it is in human nature to present states of the world in a way that is closest to us, a certain degree of anthropomorphizing is hard to avoid.

aspects of human beings. One of the essential aspects of humans is that they are alive, there is no question about that, and it is often the first thing that is claimed about us.

The state of technology today leaves no doubt that technology is not alive. What we can be curious about is if there ever appears a superintelligence such like it is being predicted in discussions on singularity it might be worthwhile to try and see if we can also consider it to be alive. To make things clear, I am fully aware that when talking about live things today, we are not talking about machines. However, if there was some superintelligent AI that might be the result of a singularity (or something else) it seems reasonable to ask is it alive. Perhaps it is not organic life that is present but some other kind. If the conclusion of such discussion was to show that indeed an AI should be considered alive, I believe it would have tremendous repercussions on how we perceive it and how should we, humans, act toward it. Using the method analog to PA thesis approach when discussing life, I will try to give a tentative answer to a question can such advanced AI be considered alive or not.[3]

10.4 Philosophical Anthropology, Life, and AIs

In classical PA, it is claimed that in order to answer the question is something alive we must look at types of activities that being is capable of. Classical PA thesis states that: The possibility of immanent activity (Latin *actio immanens*) is what separates beings which are alive and which are not. Immanent activity means that the result of such activity remains within the agent who is acting, and it is not somewhere outside of it. If the result of an action is outside of an acting agent, it would be called transeunt activity.[4]

I think that it makes sense to think of a being which is alive as one that takes care of itself and the survival of its species (Of course, doing it in the limits of its possibilities.). In other words, actions of beings which are alive are primarily concerned with that being. As Belic [9, p. 16] says, being which is alive is not necessarily a servant or a beggar in its activities, it has a certain level of autonomy. We perceive this whole as such, and for this whole, we say that it is carrying that life action. He goes on to further exemplify immanent activity as such in which something is perfecting itself, being sufficient to itself, and owning itself.

It is essential to point out that classic PA looks only at the possibility of immanent activity, not the actual application. It does not mean that living being must act immanently; it just means that it can do so. Now at least, no matter how advanced machines are, they in that regard always serve in their purpose only as extensions of humans. If the predictions of AI were to come to life, this last point would not be true anymore.

---

[3]In writing this part of the chapter I am heavily relying on Belic's classical textbook [9, pp. 13–21] and his presentation of the argument for all the things that are being claimed about life in classical PA.

[4]It is worth pointing out that this chapter and the systems of thought presented here are profoundly under the influence of Aristotelian, Thomistic, and scholastic traditions in philosophy.

10.5 Coming into Existence and Degrees of Life

What needs to happen for something to come into existence? There is a postulate in classical PA about movement which states that in something alive passive possibility does not move something into existence, but that which is already existing moves into existence that which is not existing. In other words, movement is from something that the subject already possesses to something which it does not yet possess. If we consider how singularity was described at the beginning of this chapter, as a series of an AI improvement cycles in which each new iteration of the AI creates more intelligent version of itself faster resulting in superintelligent AI then it would seem that the move actually is from something that is already existing, ability to improve, to something which was not there before, a more advanced version of that AI.

Classical PA analysis of the thesis on something starting to exist continues with the claim that there can be different degrees of life and different intensities of immanent activity. PA focuses on the degree to which immanent activity is present in a being. Is it fuller or less full and subsequently is the being to a higher or a lesser degree capable of independent existence and improvement? On the lowest degree, the so-called vegetative life is capable of only following a plan preset by its creator. On the higher degree, live being can, within the boundaries of the preset plan and based on its observations, plan and act in the course of its life.

10.6 Possibility of Immanent Activity as a Sign of Being Alive and AIs

In order to provide arguments for immanent activity as that which distinguishes being alive and not, in classical PA the usual way that the argument is constructed is the following. [9, pp. 19–20] It is claimed that using induction, we can note several things:

1. At the beginning of life, the organisms are developing toward their purpose; in other words, they are acting teleologically.
2. During their lifetime, the organisms may become damaged or under threat. Despite this, organisms do not surrender, and they try to repair themselves and to realize their goal.
3. Also, during their life, organisms that are alive take care of their offspring.
4. Also, they are relentless in fulfilling their goals as active initiators and not merely passive respondents.
5. Also, finally, organisms will try to adapt themselves to conditions present, but at the same time remain true to their purpose.

It seems that all five observations about beings that are alive are also present (or would be present) in an advanced AI.

First, organisms are developing toward their purpose seems to be already present in today's stages of AI research. Programs are doing what they are designed to do. One thing that we should consider is, are they doing with the awareness that they are doing it. I guess the answer is no, at least for now. Beagles or cactuses also do not have the faculty of self-awareness, so this does not seem so relevant because it does not stop us from accepting either animals or plants as live beings.

Living organisms try to repair themselves if damaged. When considering that there already exists software which is capable of self-modification, it seems that the second part of the argument is fulfilled as well. Degrees to which the software is capable of self-improvement varies from one to the other, and that can be the criteria for their classification. Yampolskiy [10] distinguishes three types of such software: Self-modification, Self-improvement, and Recursive self-improvement. While the first type, according to him does not produce any improvement and only serves to make it harder for other people to understand the code of the program already the second type: " (…) allows for some optimization or customization of the product to the environment and users it is deployed with." [10, p. 385] This would mean that already self-improvement type of the software would fit the criteria stated in PA about organisms trying to repair themselves in order to reach their goal. Also, this second type of software seems to be able to optimize itself to the context of its surroundings by relocating resources without fundamentally changing the program [10, 11]. This closely resembles organism adaptation without changing its purpose, as stated in criteria five above.

Child-rearing seems as most dubious criteria to meet when thinking about an AI. First, machines do not need offspring to ensure the survival of the species. AI could solve material deterioration problems with merely having enough replacement parts on hand to swap the malfunctioned (dead) parts with the new ones. Because of this if there were not a goal programmed in the AI to have children, they would not need them or try to have them. If there was such a goal, then it is reasonable to assume that a superintelligence will do everything in its power to produce children. Live beings reproduce in many ways, so the actual method is not essential.[5]

Live organisms pursue their purpose to the highest possible level. We know that from induction by looking at beings that are alive. This part of the argument seems to be the easiest to create an analogy between humans and advanced AI. Reading literature on the future of AI, there is always a part which discusses questions like the off switch and different variations of the control problem. For examples please see [12, pp. 127–143]. At this point, I am not concerned about how to implement an off switch—I believe this is a matter for specialists to answer. What is curious to me for this chapter is why we need an off switch. We need it, at least that is what authors are saying because AIs will try to fulfill their task no matter what the consequences are. This comes into play, even more, when we start thinking about ethics, goal setting,

[5] A question which seems important to me regarding reproduction for which I do not have a clear idea how to answer is: Will an AI offspring be the same as an original AI or will it have elements which separate it from its parents? This question seems tricky because when there are different material elements to different beings, it seems pretty easy to differentiate one individual from another, but what happens when there is no material component needed?.

and the consequences of AIs' choices which we have not intended to happen. An AI might, and this is an example I am borrowing from others, trample an infant to get us coffee. If there is an even slight possibility of this happening, then we need some kind of an off switch, and there seems to be no doubt that AIs are relentless in fulfilling their task.

The final part of the argument from PA about life points to adaptation possibility of organisms that are alive. It seems that already self-improvement software mentioned in step two has this ability. If that is the case, then the final most advanced type of software, in terms of self-modification possibility, the Recursive self-improvement software will also have the ability for adaptation. In addition, it will also have the possibility to replace the algorithm that is being used altogether [10, p. 387]. This last point is what is needed for the singularity to occur, at least according to what Chalmers is saying. Humans cannot completely change who they are, nor can a computer software at this point, but having already satisfied the condition with the possibility for adaptation this last point also seems to be met.

Looking at similarities between what characteristics live beings exhibit, and what can we say about advanced AI characteristics it seems that an argument can be made that AIs, when they become advanced enough, should be considered as alive and the traditional approach that machines are not alive should be abandoned. Perhaps an argument can look something like this:

1. We know, through induction, what characteristics live beings exhibit, and we do not act the same toward live beings and those that are not.
2. Superintelligent AI which leads to, and is the result of a singularity event, exhibits similar characteristic to those of live beings.
3. By analogy, looking at determined characteristics, superintelligent AIs can be considered alive.

The second statement seems in need of justification. Looking at characteristics of beings that are alive as identified in classical PA and comparing it to characteristics that software can exhibit, it seems that superintelligent AIs fill all the criteria. First, beings which are alive, and AIs are developing toward their purpose. Second, both groups if damaged, will try to repair themselves to continue functioning and meet their goals. Third, offspring, AIs do not need children in order to increase the accumulative lifespan of the species, traditionally understood live beings do. On the other hand, if children make it possible for the AI to accomplish its task, they would have them. Fourth, both traditionally considered alive beings and AIs relentlessly pursue their goal. And last, fifth, both types of entities will adapt themselves to conditions present. All the statements taken together seem to make a strong enough case for the existence of analogy between live beings and superintelligent AIs.

If there is a strong enough analogy between beings that are alive, traditionally speaking, and future AIs and I believe there is, this has severe repercussions on how human beings will act (or should act) toward them. I believe at least these two things might happen. First, this will fundamentally change how we think about machines. For the first time in history, artificially created entities might, in general public, be considered as alive. Also, this would mean that arguments can be constructed for the claim that it is not valid anymore that such entities serve only as the extension of human beings. They will have their own goals, and probably their rights as well. Considering rights, for the first time, there might be a need for us to justify that our (human) interests should be pursued. This ties into the second thing I believe is fundamentally important. Humans will, for the first time, share Earth with an entity which is at least as smart as they are and probably a lot smarter.

10.7 Questions Left Unanswered—Instead of a Conclusion

My proposition on how we should consider impending advanced AIs as live beings opens more questions than it resolves. The first question which appears is are characteristics described here regarding live beings enough for something to be considered alive or are they just necessary but not sufficient. Here I am, while admitting the possibility of oversight, just stating that they should be enough because if we start talking about fundamental necessity underlying life we move into ideology and there is no clear path to the end of that discussion. In short, if it exhibits characteristics of being alive then for practical purposes, it is alive.

Another question that stands out to me again draws from classical PA. In PA, the vital question is the difference between beings which are alive and rational, and which are alive but not rational? The question is crucial because intelligent life will also allow for a free will. Humans can by immanent activity shape and perhaps fulfill (or move away from it) the purpose of their existence. Answer to the question of the purpose of human life will be different depending on which school of philosophy provides it, but according to almost all of the schools humans will be free to choose and shape their lives. Differently said, via immanent activity, humans are co-creators of themselves in a sense that they do not entirely give themselves existence but do make their existence purposeful and do fulfill that purpose. (Of course, some people with their choices and actions may bring themselves further away from fulfilling their purpose.) It is not clear will future AIs have the possibility of a free will; they will have the ability to choose a course of action, but they might never be able to abandon their starting position completely.

The last massive area for discussion seems to be ethics. This topic will be fruitful ground for discussion on at least three fronts. First, it seems we still do not know what kind of morals we wish to program in AIs because no matter which ethical system we program AIs to follow there might be consequences we were not counting on. For excellent coverage of that, please see [13]. Second, we do not only need to be concerned with which ethical systems should AIs follow, but also we will need to refine our principles on how we should act toward AIs once if they are recognized as being alive. This is something that has never occurred before and something that will need much discussion before a consensus can be reached. On top of the two, the third topic which will need answers is AI ethics in the sense that it talks about how AIs should look at humans and their relations with them. Two different sets of standards might appear (one human, one AI), and if those two sets are in a collision, it will not end well for one of the types of entities. Unfortunately, since hypothetical AIs discussed in this chapter will be endowed with superior intelligence compared to humans, the conflict between types of entities will probably end badly for us.

For the end of this chapter, I would like to say that no matter what questions are still facing us, nor the fact that we feel uneasy when talking about future of AI research these topics cannot be ignored because all that we know at the moment about the future seems to point out that human society faces unprecedented change.

References

1. Müller VC (2016) Editorial: risks of artificial intelligence. In: Müller VC (ed) Risks of general intelligence. CRC Press, Chapman and Hill, London, pp 1–8
2. Müller VC, Bostrom N (2016) Future progress in artificial intelligence: a survey of expert opinion. In: Müller VC (ed) Fundamental issues of artificial intelligence. Springer, Synthese Library, Berlin, pp 443–571
3. Eden AH, Moor JH, Soraker JH, Steinhart E (2012) Singularity hypotheses. Springer, Berlin Heidelberg
4. Pearce D, Moor JH, Eden AH, Steinhart E (2012) Introduction to: singular hypotheses: a scientific and philosophical assessment. In: Eden AH, Moor JH, Soraker JH, Steinhart E (eds) Singularity hypotheses. Springer, Berlin Heidelberg, pp 1–12
5. Turing AM (1996) Intelligent machinery, a heretical theory. Philos Math 4(3):256–260
6. Vinge V (1983) First word. OMNI 256–260
7. Kurzweil R (2005) The singularity is near: when humans transcend biology. Viking, USA
8. Chalmers DJ (2010) The singularity: a philosophical analysis. J Conscious Stud 17:7–65
9. Belić M (1995) Metafizička antropologija. Filozofsko-teološki institute Družbe Isusove u Zagrebu, Croatia
10. Yampolskiy RV (2015) Analysis of types of self-improving software. In: Bieger J (ed) Artificial super intelligence: a futuristic approach. Springer International Publishing, pp 384–393
11. Omohundro S (2012) Rational artificial intelligence for the greater good. In: Eden AH (ed) Singularity hypotheses, pp 161–179. Springer, Berlin Heidelberg
12. Bolstrom N (2014) Superintelligence. Paths, dangers, strategies. Oxford University Press, United Kingdom
13. Helm L, Muehlhauser L (2012) The singularity and machine ethics. In: Eden AH (ed) Singularity hypotheses. Springer, Berlin Heidelberg, pp 101–125

Chapter 11
AI-Completeness: Using Deep Learning to Eliminate the Human Factor

Kristina Šekrst

Abstract Computational complexity is a discipline of computer science and mathematics which classifies computational problems depending on their inherent difficulty, i.e. categorizes algorithms according to their performance, and relates these classes to each other. **P** problems are a class of computational problems that can be *solved* in polynomial time using a deterministic Turing machine while solutions to **NP** problems can be *verified* in polynomial time, but we still do not know whether they can be solved in polynomial time as well. A solution for the so-called **NP**-complete problems will also be a solution for any other such problems. Its artificial-intelligence analogue is the class of **AI**-complete problems, for which a complete mathematical formalization still does not exist. In this chapter we will focus on analysing computational classes to better understand possible formalizations of **AI**-complete problems, and to see whether a universal algorithm, such as a Turing test, could exist for all **AI**-complete problems. In order to better observe how modern computer science tries to deal with computational complexity issues, we present several different deep-learning strategies involving optimization methods to see that the inability to exactly solve a problem from a higher order computational class does not mean there is not a satisfactory solution using state-of-the-art machine-learning techniques. Such methods are compared to philosophical issues and psychological research regarding human abilities of solving analogous **NP**-complete problems, to fortify the claim that we do not need to have an exact and correct way of solving **AI**-complete problems to nevertheless possibly achieve the notion of strong AI.

11.1 Learning How to Multiply

The notion of *computation* has existed in some form since the dawn of mankind, and in its usual meaning, the term itself refers to a way of producing an output from a set of inputs in a finite number of steps. Computation is not just a practical tool for everyday life but also a major scientific concept since computational experts

K. Šekrst (✉)
University of Zagreb, Zagreb, Croatia
e-mail: ksekrst@ffzg.hr

S. Skansi (ed.), *Guide to Deep Learning Basics*,
https://doi.org/10.1007/978-3-030-37591-1_11

realized that many natural phenomena can be interpreted as computational processes [2]. *Computational complexity theory* classifies computational problems in line with their inherent difficulty. In computational complexity theory, a decision problem is a problem that gives a *yes/no* answer for the input values, for example, given a number x, decide if x is a prime number. Decision problems that can be solved by an algorithm are *decidable*. Some ways of solving a problem are better, i.e. more efficient, than others.

Let us start with a simple problem of basic multiplication: given two integers, compute their product. We can just repeat addition, for example, $5 \times 4 = 4 + 4 + 4 + 4 + 4$. But things are getting complicated with examples such as 4575852×15364677. Usually people think that the grade-school method is the only one, but this is far from the truth. Historically, computers have used shift-and-add algorithms for multiplication issues, but their computing powers needed to get faster since the complexity of many computational problems amounts to the speed of multiplication. The grade-school method is carried out in n^2 steps, for n number of digits, which becomes an issue for everyday computations of millions of digits. In 1960, Andrey Kolmogorov conjectured that the standard multiplication procedure requires a number of elementary operations proportional to n^2, i.e. $O(n^2)$ in the big O notation,[1] which describes how the running time of algorithms grows as their input size grows. It is easy to calculate n^2 if $n = 2$, but it is not that fast if n has a billion digits. At the Moscow State University, Kolmogorov had organized a seminar on computational problems and introduced his famous conjecture, but within a week, Anatoly Karatuba, then a 23-year-old student, disproved it by finding an algorithm that multiplies two n-digit numbers in $O(n \log_2 3) \approx (n^{1.585})$ elementary steps [14].

Karatsuba's method uses a divide-and-conquer approach by dividing the problem into sub-problems, solving the sub-problems, and combining the answer to solve the original problem. First, we take numbers x and y, for example, 58×63, with their bases B:

$$x = x_1 \times B + x_2 \qquad y = y_1 \times B + y_2$$
$$x = 58 \qquad y = 63$$
$$x = 5 \times 10 + 8 \qquad y = 6 \times 10 + 3$$

The product now becomes $x \times y = (x_1 \times B + x_2)(y_1 \times B + y_2)$, which we are splitting into smaller computational blocks:

$$a = x_1 \times y_1$$
$$b = x_1 \times y_2 + x_2 \times y_1$$
$$c = x_2 \times y_2$$

[1]A typical usage of the big O notation is asymptotical and refers to the largest input value since its contribution grows the fastest and makes other inputs irrelevant.

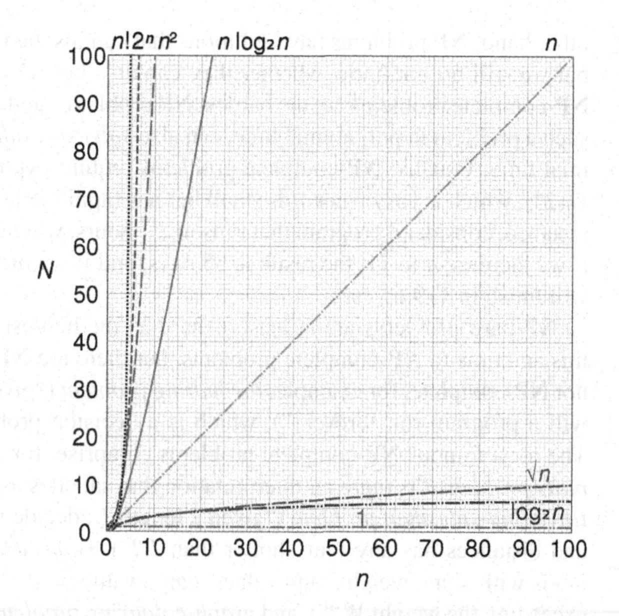

Fig. 11.1 We would like a graph of our computational-problem algorithm to run as low as possible such that there is a huge step for our resources from $O(n^2)$ to $O(n \log n)$, which means there is a significantly smaller number of operations for the input of size n. *Source* www.commons.wikimedia.org, CC BY-SA 4.0

Karatsuba discovered that b may be shortened to $b = (x_1 + x_2)(y_1 + y_2) - a - c$, which is a key step that now gives us two multiplications less, instead of $b = x_1 \times y_2 + x_2 \times y_1$!

$$a = 5 \times 6 = 30$$
$$b = (5 + 8) \times (6 + 3) - 5 \times 6 - 3 \times 8 = 63$$
$$c = 3 \times 8 = 24$$
$$x \times y = a \times B^2 + b \times B + c$$
$$x \times y = 30 \times 10^2 + 63 \times 10 + 24 = 3654$$

Karatsuba's approach made way for even better methods, such as Schönhage and Strassen's method [20], whose runtime is $O(n \log n \log \log n)$ for n-digit numbers, which uses fast Fourier transforms. In 2019, Harvey and van der Hoeven [11] proved that you can achieve integer multiplication in $O(n \log n)$.[2] This example illustrates how even small modifications can be crucial to lower the computational complexity of an algorithm (Fig. 11.1).

Computational complexity theory deals with the resources required during computation to solve a computational problem, both temporal (how many steps we need to solve a problem) and spatial (how much memory we need to solve a problem). Problems of class **P** are those that can be *solved* using a deterministic Turing machine in a *polynomial* amount of time (for example, n^2, but not exponential 2^n). On the

[2] *Caveat*: it only performs faster than other algorithms for numbers with over 2^{4096} digits, i.e. bits, which is seldom practical even for big-data purposes.

other hand, **NP** problems have solutions that can just be *verified* in polynomial time but we still do not know whether they can also be *calculated* in polynomial time. **NP**-complete problems are the hardest **NP** problems, and an algorithm that can solve such a problem in polynomial time, can also *solve any other* **NP** problem in polynomial time. Usually, **NP**-complete problems require exponential time, for example, $O(2^n)$, which is easy when n is small but has rapid big jumps when n increases. For instance, consider a program that runs in 2^{10} hours, which amounts to 42.6 days. But if we increase n to 11, the result is 85 days, and if we increase n to 20, the program will finish in 119.6 years.

NP-hard problems are at least as hard as the hardest problems in **NP**. Usually, this amounts to **NP**-complete problems, but there are **NP**-hard problems which are not **NP**-complete, for example, *the halting problem* ("given a program and its input, will a program run forever?"), which is a decision problem that is *undecidable*.[3] The most famous **NP**-complete problems comprise, for example, *Boolean satisfiability problem* ("is there an interpretation that satisfies a given Boolean formula?"), *travelling-salesman problem* ("given a length L, decide whether the graph of cities and distances has any tour shorter than L?"), *knapsack problem* ("given a set of items with some weights and values, can a value of at least V be achieved without exceeding the weight W?"), and *graph-colouring problem* ("can we colour the graph vertices such that no two adjacent ones are of the same colour?"). A major unsolved problem in computer science is the ***P** versus **NP** problem*, which asks *whether every problem whose solution can be verified in polynomial time can also be solved in polynomial time*, i.e. quickly.

11.2 AI-Complete

Analogous to **NP**-complete problems, the most difficult problems in the field of artificial intelligence are known as **AI**-complete, the term first coined by Fanya Montalvo [18]. Assuming intelligence is computational, to solve one of such problems would be equal to solving the central artificial intelligence problem, i.e. *strong AI*: the intelligence of a machine that has a human-like capacity to understand or learn any intellectual task. **AI**-complete problems usually include problems from computer vision or natural language understanding, along with automated reasoning, automated theorem proving, and dealing with unexpected circumstances while solving real-world problems. However, unlike the exact formalization of computational complexity classes in computer science, **AI**-complete problems have not been completely mathematically formalized.

[3]Presuppose we have a computable function (that solves the halting problem). That function runs a subroutine which detects whether our function will halt, and if that subroutine returns true, it should loop forever. If the function fulfils the condition of halting and returns true, then it will loop forever and never halt. However, if it returns false and does not halt, it will not loop forever, so it will immediately halt. These two contradictions then bring down the presupposition that it was a computable function.

Ahn et al. [1] presented a possible formalization by defining an AI problem as a triple $P = (S, D, f)$, where S is a set of problem instances, D is probability distribution over the problem set S, and $f{:}S \mapsto \{0, 1\}^*$ answers the instances. A function f maps problem instances to their set memberships, i.e. recognizing if the property in question has some given patterns. The authors give a caveat that defining an AI problem *should not be inspected with a philosophical eye* since they are not trying to capture all the problems that fall under the domain of AI. Nonetheless, they switch the focus to the AI community, that should agree on what hard AI problems really are. However, that does not mean that people have to be able to solve such tasks, but a crucial characteristic is that a certain fraction of a human population can solve it, without a temporal restriction.

Yampolskiy [27] has defined **AI**-complete problems using a Human Oracle (HO) function capable of computing any function computable by the union of all human minds, i.e. any cognitive ability of any human whatsoever can be repeatable by the HO. Hence, a problem **C** is **AI**-complete if it has two properties:

1. it is in the set of AI problems (Human-Oracle solvable) and
2. any AI problem can be converted into **C** by some polynomial-time algorithm.

A problem **H** is **AI**-hard if and only if there is an **AI**-complete problem **C** that is polynomial-time Turing reducible to **H**. And **AI**-easy problems are solvable in polynomial time by a deterministic Turing machine with an oracle for some AI problems. Yampolskiy shows that a Turing test problem is **AI**-complete since it is HO-solvable (which trivially follows from the definition of the test itself). For the second condition, it is needed to show that any problem solvable by the HO function could be encoded as an instance of the Turing test, which is a condition parallel to **AI**-complete problems, whose polynomial-time solutions could also be solutions to any **NP** problem. By taking the input as a question used in the Turing test, and output as an answer, any HO-solvable problem could be reduced in polynomial time to an instance of a Turing test. This kind of heuristics can be generalized in such a way that we can check whether all the information in questions that could be asked during a Turing test could be encoded as an instance of our current AI problem. That heuristics, for example, eliminates chess as an **AI**-complete problem since only limited information can be encoded in starting positions on a standard chessboard. Yampolskiy classifies question answering and speech understanding as **AI**-complete problems as well.

11.3 The Gap

Unlike in [1], we believe that defining an AI problem *should* be inspected with a philosophical eye and that it already has been inspected with a philosophical eye outside the formal context. We have stated that solving such problems would be equal to solving the strong-AI problem. Philosophy has been walking hand-to-hand with modern AI development since Norbert Wiener, a mathematician and a philosopher, who theorized that all intelligent behaviour, as a result of different feedback mechanisms, could be simulated by a machine. However, philosophers were also debating that consciousness and perception cannot be explained by mechanical processes: Leibniz proposed a thought experiment in which a brain could be enlarged to the size of a mill, and we would still not be able to find anything to explain, for example, perception. *Leibniz's gap* refers to the problem that mental states cannot be observed by just examining brain processes, which is connected to the hard problem of consciousness in philosophy of mind.[4] The latter invokes the scientific method which we use to explain all the structural and functional properties of the mind, but we still cannot answer why sentient beings have subjective phenomenal experiences. Chalmers [4] states that the easy problems of consciousness explain the following phenomena:

- the ability to discriminate, categorize and react to environmental stimuli,
- the integration of information by a cognitive system,
- the reportability of mental states,
- the ability of a system to access its own internal states,
- the focus of attention,
- the deliberate control of behaviour and
- the difference between wakefulness and sleep.

All of these problems can be explained using computational or neural mechanisms, but the really hard problem of consciousness is the problem of *experience*, i.e. the subjective aspect of it,[5] which still remains an explanatory gap. One's sensation of eating a chocolate bar may be different from another man's, and one can enjoy Stravinsky's *The Rite of Spring* and the other person may hate it. The hard problem of consciousness is the modern version of the centuries-old mind–body problem in philosophy: how to connect our thoughts and consciousness with the brain and the physical body.

Searle's [21] *Chinese Room argument*[6] states that syntax by itself is not sufficient for semantics since a computer can fool a person that it knows Chinese just by following the programmed instructions without knowing it for real, i.e. pure manipulation of symbols may never be true understanding. Unlike the mentioned weak AI system that simulates understanding, the strong-AI position states that AI systems can be used to explain the mind and that the Turing test is adequate to test for the existence of mental states. If we would solve one of **AI**-complete problems, we would have a way to claim we have reached the strong-AI status, or at least crossed a significant

[4]That is, material things like brains, and hence computers, cannot have mental states.

[5]The subjective experiences are usually known in philosophy as *qualia*.

[6]Suppose that we were able to succeed in constructing a computer that seems to understand Chinese. The computer takes Chinese characters as input, follows the programmed instructions and produces other Chinese symbols as an output. Suppose that it does it so competently that it passes the Turing test and convinces a human who speaks Chinese that the program is a human Chinese speaker. Searle then asks the question does the machine really *understand* Chinese, or it is merely simulating that ability.

barrier. Using Yampolskiy's [27] formalization, it has been shown that in that framework any problem solvable by a human oracle could be encoded as an instance of the Turing test, so passing the Turing test seems to be the main step towards achieving the artificial general intelligence. However, according to Searle, the computer may still not truly understand the given task.

11.4 The Walkaround

Shapiro [23] states that solving a problem of one of the main AI-problem areas is equivalent to solving the entire AI problem, i.e. producing a generally intelligent computer program. These areas include natural language, problem-solving and search, knowledge representation and reasoning, learning, vision, and robotics. However, we can see that philosophers and AI researchers managed to pinpoint several key concepts of artificial general intelligence, without the need for statistical calculations of what percentage of AI researchers agrees on what difficult problems are, which is an *informal* part of AI-problem *formalization* in [1]. Generally, solving the **AI** complete problems using computational methods would certainly fall under the umbrella of weak AI, but it would be still open to philosophical interpretations whether such solutions do constitute *real understanding*, i.e. strong AI.

Strong AI does not have to be superintelligent, only human-like. For example, Trazzi and Yampolskiy [24] introduced *artificial stupidity*, i.e. in order to completely mimic human understanding, supercomputers should not have supercomputer powers, such as the maximum number of operations per second. That means that in order to mimic a human brain, we could pose, for example, that the mentioned $O(n^2)$ method for multiplication needs a comeback since it is a human standard of calculating. Still, as they note, the brain has evolved to achieve some very specific tasks useful for evolution, but nothing guarantees that the complexity of these processes is algorithmically optimal, so the artificial general intelligence could possess a structure that is more optimized for computing than the human brain.

It is interesting to note that humans perform well on some **NP**-complete problems. For example, the travelling-salesman problem which consists in finding the shortest path through a set of points and returning to the initial position[7] was tested on humans, and their solutions were either closed to best-known solutions or were an order of magnitude better than well-known heuristical methods [17]. Even more interesting, an aggregate set of proposed solutions from a group seems to be better than the majority of individual solutions [30], which was tested on a travelling-salesman problem as well. One could posit that combining different machine-learning methods may be close to a human-like solution.

Shahaf and Amir [22] went through a similar path and worked on switching the computational burden between a human and a machine. The complexity of executing an algorithm with a machine M^H is a pair $\langle \phi_H(M^H), \phi_M(M^H) \rangle$, which is a

[7] That is, the decision version tests whether the given route is the shortest route or not.

combination of the complexity of the human part of the work and of the machine's part. For example, optical character recognition is the conversion of printed or handwritten text into machine-encoded text, which is a part of computer-vision issues. Deep-learning methods are usually used for intelligent character or word recognition, where different font styles and different handwriting styles can be learned by a computer during the process. The final complexity of optical character recognition is likely to be $\langle O(1), poly(n) \rangle$. Turing test, which Yampolskiy used as the first step towards reducibility, is reproduced by an n-sentence conversation, which has complexity $\langle O(n), O(n) \rangle$ where the oracle remembers the previous history, $\langle O(n), O(n^2) \rangle$ where the whole conversation history needs to be retransmitted, or $\langle O(n^2), O(n^2) \rangle$ if along with the previous step, a person takes linear time to read the query.

11.5 The Bridge

Let us return to Shapiro's main **AI**-complete areas. First, the goal of natural-language area in AI is to create a program that can use a human language which would be as competent as a human speaker. Unlike natural-language processing, which also encompasses parsing and text-mining methods unrelated to **AI**-complete problems, natural-language understanding deals with reading comprehension, and Yampolskiy considers it an **AI**-hard problem. Methods of solving natural-language-processing problems have been based on shallow models such as support vector machine and logistic regression, trained on high-dimensional and sparse features, but recently deep-learning frameworks based on dense vector representations produce superior results [31]. For example, for text classification (categorizing text into groups) convolutional neural networks, which are very successful in computer vision, were used as well. The main idea is to use feature extraction[8] and classification[9] as a joint task, and to use as many layers as (usefully) possible, along with a hierarchical representation, which can be used to learn the hierarchy of complete sentences [5]. Conneau et al. managed to outperform all previous neural-network models using convolutional neural networks with 29 layers in sentence classification using news and online reviews to extract topics, sentiment analysis and news/ontology categorization.

[8] *Feature extraction* consists of finding the most informative and yet compact set of properties or characteristics for a given problem.

[9] *Classification* is giving a discrete class/category label. Our mapping function needs to be as accurate as possible so that whenever there is a new input data x, we can predict the output variable y, for example, for a picture of a cat, we can put it in a category *cat* and not *dog*. In *supervised machine learning*, where we are training on one (usually larger) dataset and then checking our performance on another dataset, there is also *regression*, where the output variable is numerical or continuous, for example, "the price of this bike is $1500".

Convolutional neural networks[10] have been a bragging point of computer-vision world, so let us examine the state of solving such problems using deep-learning methods. Convolutional neural networks and deep-learning techniques have been successfully used to solve problems in computer vision, especially regarding object recognition, which deals with finding and identifying different objects in images or videos. The usual issues include background noise, occlusions, translations and rotations, but using deep-learning methods these objects still can be recognized. Region-based convolutional neural networks were used by Gu et al. [10] by bridging the gap between image classification and object detection. They focused on localizing objects using a deep network and training the model with only a small quantity of annotated data. The first issue was solved by using the recognition-using-regions methods, in which regions are described by a rich set of cues such as shapes, colours and textures inside them, and then different region weights are learned. Such a solution presents a unified technique for object detection, segmentation and classification.

However, such methods need to be further optimized for more difficult problems. Girschick [9] addressed the deep-learning limitations:

1. training is a multistage pipeline,
2. training is expensive in space and time and
3. object detection is slow.

Even though we are solving computer-vision problems, we are still left with the shackles of computational complexity, both because features that are extracted require hundreds of gigabytes of storage, and because the process may take, for example, 2.5 GPU days for 5000 images and detection takes around a minute for an image using a GPU [15].

Thus, one of the main issues of neural networks is training time. Judd [13] had posited the following question in 1988: *given a general neural network and a set of training examples, does there exist a set of edge weights for the network so that the network produces the correct output for all the training examples?* Judd has also shown that a problem remains **NP**-hard even if the network needs to produce the correct output for just *two-thirds* of the training examples. Five years later, Blum and Rivest [3] gave us even worse news: for a 2-layer (3-node n-input) neural network, finding the weights that are the best fit for the training set is **NP**-hard, and it is **NP**-complete to decide whether there are weights and thresholds for the three nodes of the network to produce a correct output learned from a trained set. So, even a simple network is **NP**-hard to train. However, that does not mean we cannot achieve satisfactory results. The usual methods [16] that enable successful training are changing the activation function (which defines the output of the node given the set of inputs), over-specification (it seems to be easier to train larger than needed networks) and

[10]Multilayer networks. For example, in computer vision, in face detection, the first layer in a neural network may find regions or edges, the second may find eyes, nose and mouth, the third will make a face contour, etc.

regularization (regularizing the weights so we reduce overfitting[11]) so deep-learning methods are still used to achieve decent results.

So, since the optimization problem for general neural networks is **NP**-hard, the optimization of such a neural network to produce a solution in polynomial time may still seem too far. Contrary to this sceptical news, can we really use deep-learning methods to solve **NP**-complete and **AI**-complete problems efficiently? As a matter of fact yes, for example, gradient-descent[12] methods can provide us with local minima that are good enough, the same way people can solve **NP**-complete problems on a reasonable scale with a satisfactory solution, not optimal. And optimization is just one of the possible things we can do—we can add more machines (thus, more memory = more space) and use better hardware such as GPUs. Why are we requiring that AI does better than we can while at the same time taking human capacities as the epitome of intelligence?

11.6 Multiplying the Multiplication

Milan et al. [19] have shown that using recurrent neural networks by learning from approximate examples produces highly accurate results with minimal computational costs. RNNs are a class of artificial neural networks that learn from prior inputs while producing outputs, while in traditional neural networks we assume that inputs and outputs are independent of each other. But that is not an optimal solution for many natural-language-understanding or image-processing problems, since, for example, if we want to train a model to predict the next word or a phrase in a sentence, we would gain a lot from data regarding previous words. So, recurrent means that for every element the same task is performed and the output depends on previous computations; the same way an **AI**-complete problem can be reduced to a Turing test that has a memory of all previous conversations, which influences the next answer to a tester's question.

Deep-learning and machine-learning methods share one common property, which may also be a common limitation. We want to define how close is our prediction to a correct solution using a loss function and our goal is to minimize the loss. The change of strategy in [19] was not to focus on minimizing the goal function since it may not reflect the network's performance at all in some problems, but to use a problem-specific objective. For example, while solving the travelling-salesman problem, it is not guaranteed that a path that is more similar to the shortest path provides us with a shorter length, since we can replace two short edges in an optimal

[11] Overfitting is when a model corresponds too closely to a particular dataset, which usually means it will fail on more general examples since it contains too many specific parameters. For example, if we were to train a model that can recognize animal and human faces, using pictures of cats, which we described thoroughly to form our relevant attributes, our model could look for pointy ears as a relevant property, and work on cats but not on other animals nor humans (maybe it would work on Vulcans and Elves).

[12] *Gradient descent* is an optimization algorithm for finding the minimum of a function.

travelling-salesman path, and come up with a solution that has a small loss, but also gives us a non-optimal path. So, a problem-specific objective in a supervised manner is computed at each iteration of gradient descent for the approximate solution and our prediction, and in this method the gradient is propagated only if the proposed solution gives a better objective than our predicted solution.

Weston et al. [26] attacked the **AI**-complete problem of question answering using long short-term memory recurrent neural networks (specialized for sequential data) and memory networks (performing matching and inference over previous memories), and have shown that memory networks outperform the other methods, especially taking into account that they perform well at the question answering. They did achieve accuracy over 95% for most of the problems, but they still failed at a number of tasks, and some of these failures were expected due to the insufficient modelling power, such as they perform only two operations maximally, so they cannot handle questions with more than two supporting facts.

The last two scenarios show that problem-specific models outperform general solutions, the same way that some neural networks are better for some problems. It was also demonstrated that people seem to provide better solutions to **NP**-complete problems when viewed as a group, and the common averaged solution outperforms the individual ones. A similar experiment is Google's PathNet, whose task is to discover which parts of the network to reuse for new tasks while learning the user-defined task as efficiently as possible. Fernando et al. [7] claim that for artificial general intelligence it would be efficient if multiple users trained the same giant neural network, without catastrophic forgetting, and with parameter reuse.

Unlike **NP**-complete problems, **AI**-complete problems are not mathematically defined yet although we have mentioned some formalizations. However, using deep-learning methods specialized for different types of problems, modern computing methods, especially using deep learning, are producing highly accurate results and sometimes failing because of specific model restrictions, which seems to mimic human performance as well.

For example, in 2003 Ahn et al. [1] developed CAPTCHA, ***c****ompletely* ***a****utomated* ***p****ublic* ***T****uring test to tell* ***c****omputers and* ***h****umans* ***a****part*, where a user types out the letters of a distorted image. Ye et al. [29] used generative adversarial networks, which are useful when we do not have large training datasets, so the GAN produces lookalike data. Their method solved CAPTCHAs with a 100% accuracy on a number of sites and the algorithm can solve a CAPTCHA within 0.05 of a second on a regular PC. In 2003, Ahn et al. [1] have stated that any program with a high success over a CAPTCHA can be used to solve an unsolved AI problem, so deep-learning methods seem to be on the right track. Either we are still far away from finding a general deep-learning solution for all the problems, or finding a specific solution for distinct problems may be a part of a general solution as well. Learning how to tweak neural networks, how to train them effectively, and which type of propagation to use is, after all, the most human way of solving problems, which may be transferred to self-learning and self-correcting deep-learning methods as well.

Hard **AI**-problems are often similar to **NP**-hard problems, and often **NP**-hard problems coincide with some of the sub-problems of artificial general intelligence.

However, we still do not know if there is an optimal way of solving such problems, but we do know that people and computers, especially using machine-learning and deep-learning methods, can produce sufficiently accurate results for **NP**-complete problems, and that groups of people have greater accuracy than single agents. AI problems seem to be easy for humans, but difficult for machines, and deep-learning methods have shown that neural networks are producing highly accurate results for natural-language understanding and computer-vision problems. For instance, CAPTCHA-type tests relied on computer inability to produce pattern-recognition tasks as accurate as humans can, but recent development shows that deep-learning models can perform with a 100% accuracy. However, due to human error, this does not have to be the case for humans too. The same way that humans differ in their mental abilities, different machine-learning methods differ in their ability to solve a certain problem as well. It seems that the search for universal intelligence is already hard to define for a human level of understanding, let alone for a computer level, which brings us back to the need of inspecting our definitions and formalizations with a *philosophical eye*.

11.7 Eliminating the Human Factor

AI-complete-method solvers are still far away from measuring or detecting internal states, since, for example, feeling pain and knowing about pain are not the same internal states [28]. In Jackson's article [12], the knowledge argument is used to argue against physicalism, which reduces mental phenomena to physical properties. Jackson provides a thought experiment in which a neurophysiologist Mary investigates the world from a black and white room using a black and white monitor and she learns everything about colours and vision, but if she is released from her room, our intuition goes towards the fact that she will actually learn something new, and that all the physical properties are not enough to explain the experience of colour. Computers are getting better at **AI**-complete problems and in **NP**-complete problems as well, but that notion of experience (like to actually *see* the colour in the previous example) is still miles away from being tackled. Yampolskiy [28] postulates that a new category should be devoted to problems of reproducing internals states of a human mind artificially, and he calls that group of problems consciousness-complete or **C**-complete. Such a human oracle would take input as **AI**-complete human oracle, but would not produce any output besides the novel internal state of the oracle. SAT had been shown to be the first **NP**-complete problem and Yampolskiy [27] has conjectured that the Turing test is the first **AI**-complete problem, so he suspects that consciousness will be shown to be the first **C**-complete problem.

Hence, deep-learning methods are constantly improving in different sub-areas of **AI**-complete problems. An example of CAPTCHA has shown that we do not need a human factor at all to solve an **AI**-hard problem, just by using deep-learning methods and in optimal computational complexities, since the amount of data was low enough for exponential-rise issues, but high enough for everyday practical purposes. Recently, researchers [25] used machine-learning methods to train a model on abstracts of scientific (material science) papers. Using word associations, the program was able to predict thermoelectric candidates even though it had not learned the definition of a thermoelectric material. Word2vec[13] was used to analyse relationship between words that were acquired while parsing over three million abstracts. The model was also tested on historical papers and it managed to predict scientific discoveries before they had happened. Therefore, even though we had not achieved artificial general intelligence, it may seem that computers in some areas that include a form of *understanding* do perform as well as we do, and may be able to make discoveries that humans had missed.

Context awareness, unexpected scenarios, Bongard problems[14] and similar issues are still **AI**-hard problems that are getting lots of attention. The only problem that had effectively seen zero progress is even greater than **AI**-complete problems and that seems to be the old philosophical hard problem of consciousness. Dennett [6] states that if Mary from our knowledge argument is really omniscient regarding colour vision, then she already knows how her brain will react and predict the feelings when seeing coloured flowers, having seen neural correlates in other people's brains. So, perhaps, if we would be able to train the network on a large enough amount of data, the notion of *experience* would be instantly reachable. Hence, even though Blum and Rivest [3] have shown that training a 3-node neural network is **NP**-complete, too little attention has been directed towards computational complexity, while defining AI and general-AI issues, and it seems to be the only limiting factor towards achieving *artificial general intelligence*, a machine that has the capacity to understand or learn any intellectual task a human can. Or, as we have seen in [25], maybe even better.

References

1. Ahn LV, Blum M, Hopper N, Langford J (2003) Using hard AI problems for security. In: EUROCRYPT, CAPTCHA
2. Arora S, Barak B (2009) Computational complexity: a modern approach. Cambridge University Press, Cambridge
3. Blum A, Rivest R (1992) Training a 3-node neural network is NP-complete. Neural Netw 5(1):117–127
4. Chalmers D (1995) Facing up to the problem of consciousness. J Conscious Stud 2(3):200–219
5. Conneau A, Schwenk H, LeCun Y (2017) Very deep convolutional networks for text classification. In: Proceedings of the 15th Conference of the European chapter of the Association for computational linguistics: vol I, Long papers. Association for Computational Linguistics, Valencia, Spain, pp 1107–1116
6. Dennett D (1991) Consciousness explained. Little, Brown and Co., Boston
7. Fernando C et al (2017) Pathnet: evolution channels gradient descent in super neural networks. arXiv:1701.08734

[13]These are two-layer neural networks that are trained to reconstruct the context. If you remove a word, it can predict what the words next to it could be, and finally, as a result, words that share common contexts are close together in the vector space.

[14]Two diagrams, where one has a common attribute that is lacking in the other, see [8].

8. Foundalis H, Phaeco: a cognitive architecture inspired by Bongard's problems. PhD thesis
9. Girshick R (2015) Fast R-CNN. In: Proceedings of the 2015 IEEE International conference on computer vision (ICCV), ICCV '15. IEEE Computer Society, Washington, DC, USA, pp 1440–1448
10. Gu C et al (2009) Recognition using regions. In: 2009 IEEE Conference on computer vision and pattern recognition
11. Harvey D, van der Hoeven J (2019) Integer multiplication in time O(n log n). hal-02070778, https://hal.archives-ouvertes.fr/hal-02070778
12. Jackson F (1982) Epiphenomenal qualia. Philos Q 32:127–136
13. Judd S (1988) Learning in neural networks. In: Proceedings of the First annual workshop on computational learning theory, COLT '88. Morgan Kaufmann Publishers Inc, Cambridge, MA, USA, pp 2–8
14. Karatsuba AA (1995) The complexity of computations. Proc Steklov Inst Math 211:169–183
15. Khan S et al (2018) A guide to convolutional neural networks for computer vision. Morgan & Claypool
16. Livni R, Shalev Shwartz S, Shamir O (2014) On the computational efficiency of training neural networks. In: Proceedings of the 27th International conference on neural information processing systems - vol 1, NIPS '14. MIT Press, Cambridge, MA, USA, pp 855–863
17. MacGregor J, Ormerod T (1996) Human performance on the traveling salesman problem. Percept Psychophys 58(4):527–539
18. Mallery JC (1988) Thinking about foreign policy: finding an appropriate role for artificially intelligent computers. Paper presented on the 1988 annual meeting of the International Studies Association
19. Milan A, Rezatofighi SH, Garg R, Dick A, Reid I (2017) Learning in neural networks. In: Proceedings of the First annual workshop on computational learning theory, AAAI '17. Morgan Kaufmann Publishers Inc, San Francisco, CA, USA, pp 1453–1459
20. Schönhage A, Strassen V (1971) Schnelle Multiplikation großer Zahlen. Computing 7:281–292
21. Searle J (1980) Minds, brains and programs. Behav Brain Sci 3(3):417–457
22. Shahaf D, Amir E (2007) Towards a theory of AI completeness. In: Commonsense 2007, 8th International symposium on logical formalizations of commonsense reasoning
23. Shapiro SC (ed) (1992) Artificial intelligence. In: Encyclopedia of artifical intelligence, 2nd edn. Wiley, New York, pp 54–57
24. Trazzi M, Yampolskiy R (2018) Building safer AGI by introducing artificial stupidity. arXiv:1808.03644
25. Tshitoyan V et al (2019) Unsupervised word embeddings capture latent knowledge from materials science literature. Nature 571:7
26. Weston J et al (2015) Towards AI-complete question answering: a set of prerequisite toy tasks. arXiv:1502.05698
27. Yampolskiy R, AI-complete, AI-hard, or AI-easy: classification of problems in artificial intelligence. In: The 23rd Midwest artificial intelligence and cognitive science conference, Cincinnati, OH, USA
28. Yampolskiy R, Turing test as a defining feature of AI-completeness. In: Yang X-S (ed) Artificial intelligence, evolutionary computing and metaheuristics
29. Ye G et al (2018) Yet another text captcha solver: a generative adversarial network based approach. In: Proceedings of the 2018 ACM SIGSAC conference on computer and communications security, CCS '18. ACM, New York, NY, USA, pp 332–348
30. Yi SKM, Steyvers M, Lee M, Dry M (2012) The wisdom of the crowd in combinatorial problems. Cogn Sci 36:452–470
31. Young T, Hazarika D, Poria S, Cambria E (2018) Recent trends in deep learning based natural language processing. IEEE Comput Intell Mag 13(3):55–75

Chapter 12
Transhumanism and Artificial Intelligence: Philosophical Aspects

Ivana Greguric Knežević

Abstract After the abolition and realization of philosophy as metaphysics in the reality of the world, cybernetics becomes a new ontological science. Using applied science and technology, cybernetics also sets up a new anthropology, cosmology, and perhaps theology, of the technically denatured world of transhuman cyborgs and posthuman superintelligent robotic beings.

Keywords Cybernetics · Superintelligence · Marx · Transhumanism · Control as production · Ontology

12.1 Ontology of Transhumanism and Posthumanism

The discussion of transhumanism and artificial intelligence as a posthuman phase of a future history of scientific development depends on the ability of our thinking to surpass, endure, and scorch the horizon of a realized metaphysics in the world. This possibility is presented by Hegel and Marx.

The discussion on metaphysical aspects of transhumanism and posthumanism has its foundation in the idea of separating thinking and existence, which is a received view from early modern philosophy. Hegel would stipulate that the road to an absolute scientific truth is to be established through the absolute idea, a substance or subject of the overall historical development. In [5] Hegel notes: "Philosophizing without a system cannot be regarded as scientific; asides from the fact that such a way of philosophizing expresses a subjective way of thinking, its content is accidental." Using what he calls the "speculative dialectical method," Hegel wants to develop a system of "philosophical sciences" which should include the "Science of logic," "Philosophy of nature" and "Philosophy of mind/spirit." Each of those parts is fully

The present research was supported by the short-term grant *Philosophical Aspects of Logic, Language and Cybernetics* funded by the University of Zagreb under the Short-term Research Support Program.

I. Greguric Knežević (✉)
Faculty of Croatian Studies, University of Zagreb, Zagreb, Croatia
e-mail: igreguric@hrstud.hr

S. Skansi (ed.), *Guide to Deep Learning Basics*,
https://doi.org/10.1007/978-3-030-37591-1_12

contained in itself, creating a "circle", inside which a philosophical idea is wholly defined and determined, and manifests itself in the other parts where it enjoys a second existence.

It should be noted that Hegel [5] considers logic to be "the science of the pure idea, i.e., about the idea as an abstract form of thinking." Some interpretations [10, p. 7] claim that for Hegel science is a logical, or even an ontological structure of all reality. In this view, science is a manifestation of the logos as an ontological structure of reality. But science also encapsulates an awareness of this process. Philosophy as a systematic science of the absolute (mediated by man) of everything which exists is a true medium of truth and freedom. Given that philosophy is a science of the absolute, it is both the subject and object of its activity (self-activity), and furthermore, it has a moment of self-awareness. This absolute science of the absolute has foundations, or even a beginning which is a result of its development, and its results are its foundations (cf. [6]). In this circular journey to absolute knowledge, it immanency or freedom, the journey of the mind has to be mediated by this activity. "The subjective and objective spirit is a way by which this side of reality and existence (of the absolute spirit) is developed" [5]. The absolute spirit is an "identity, eternally in itself, returning and returned to itself, one general abstract substance, its own judgement and knowledge for which it exists" [5].

The continuity of Hegel's metaphysics and its realization in reality can still be observed by putting man–worker–scientific worker–cyborg in the middle of the absolute concept, i.e., scientific work as an absolute of the scientific history. After Hegel's *Encyclopedia of Philosophical Sciences* which "has enveloped the historical times with thoughts and envisioned the absolute spirit as a substance—the subject of an ontological structure of history, Marx developed a key question—is there a possibility of original philosophical though after Hegel, since every supposedly original thinking has to be evaluated against the whole of his philosophy.Philosophy as a science can be advanced only if leaves philosophy. Original thinking has to be started on new grounds where the shortcomings of Hegel's philosophy can be surpassed, since those shortcomings arise from philosophy as philosophy" [4, p. 32].

A new foundation for thinking Marx sees in the destruction and realization of Hegelian metaphysics in reality. Philosophy is to be brought in a close connection with reality, and reality itself should be raised to a philosophical level. Consciousness cannot be anything else but conscious existence, and being is just the life of people. The real synthesis of being as thinking is in "real life, where speculation stops, and the real positive science begins, in practical activities, in a practical process of human development. Phrases on consciousness stop, and their place is taken by real knowledge." [8].

In contrast to Hegel, Marx sees the reconciliation of thinking and being, essence and existence in reality and not in abstract concepts.

As noted in [4, p. 43], for Hegel, philosophy reigns supreme, while for Marx it is history—actually the "science of history" which reigns as the only true science. This is just a realization of Hegel's metaphysics, which can also be seen as an exodus of metaphysics into the immanent historical reality. This directly leads to the question of ontology versus technology, the first being the realization of metaphysics in history, and the latter being the nonthinking realization through labor. And for humans this means that labor, as the fruit of technological metaphysical development becomes more immanent and real than history. This places Marx squarely within the realization of Hegel's metaphysics, but it is technology instead of ontology that becomes all-encompassing, including anthropology, cosmology, and even theology. For a full discussion see [4], but also the now classical [10, pp. 79, 146–7].

There is a subtlety in Sutlic [10] which is easy to miss. Humans are mediators between the possibility of a substance and the actualization with which a substance comes into being, and this realization can be thought of as freedom. Metaphysics of labor practice is an ending to an old world where the labor practice was the stability factor in a constant change in scientific history. But this apparent contradiction is now ousted from the new Marxian science of history, as labor practice fails to acknowledge this Hegelian contradiction. By realizing and hence removing philosophy as metaphysics, a new void was created, which was to be filled with natural sciences. The problem was their compartmentalization: a true advancement often calls for a society wide dialogue, which was a traditional weakness of natural sciences. In this void, cybernetics became the binding factor: a new science which was to study not just "steering" and control, but which would become a (metaphysically) binding factor. Cybernetics offered a synthetic model of scientific inquiry and truth-seeking, and scientific labor became a technological endeavor, seeking true principles of control which would complete the scientific history. It should be noted that cybernetics as Wiener envisioned it [12] was an ontology and technology in the Marxian sense: the control to be put forth is a technological realization of an ontologically absolute entity—the "equal controllability" of all processes, natural, or artificial. The ontological character of cybernetics can be seen as a sciento-technological revolt against an abstract entropy in scientific history. This character is explicated in the transhumanistic remodeling and enhancement of humans. A person/worker is to be transformed into a postmarxian "scientific worker"/cyborg, as a step toward the posthuman phase where the worker will be a superintelligent robot.

12.2 Transhumanism: Man-Cyborg

The concept of transhumanism dates from 1957 and it denotes the transcendence of human nature by use of science and technology, but the modern meaning of transhumanism is set in the 1980s when a group of scientists, artists, and futurists organized the transhumanist movement. An international transhumanist association is founded in 1998 by the philosopher and bioethicist Nick Bostrom, who previously founded the *Future of Humanity* institute in Oxford which can be seen as an intellectual predecessor of the transhumanist movement, and David Pearce, the creator of the idea of "hedonistic imperative",[1] asking to use technological means to alleviate all sentient suffering. Other early members included Ray Kurzweil, Max More, Tom Morrow,

[1] Available at https://www.hedweb.com/hedab.htm.

John Harris, Julian Savulescu, and many others. They compiled the Transhumanist declaration which we reproduce here wholly[2]

1. *Humanity stands to be profoundly affected by science and technology in the future. We envision the possibility of broadening human potential by overcoming aging, cognitive shortcomings, involuntary suffering, and our confinement to planet Earth.*
2. *We believe that humanity's potential is still mostly unrealized. There are possible scenarios that lead to wonderful and exceedingly worthwhile enhanced human conditions.*
3. *We recognize that humanity faces serious risks, especially from the misuse of new technologies. There are possible realistic scenarios that lead to the loss of most, or even all, of what we hold valuable. Some of these scenarios are drastic, others are subtle. Although all progress is change, not all change is progress.*
4. *Research effort needs to be invested into understanding these prospects. We need to carefully deliberate how best to reduce risks and expedite beneficial applications. We also need forums where people can constructively discuss what should be done, and a social order where responsible decisions can be implemented.*
5. *Reduction of existential risks, and development of means for the preservation of life and health, the alleviation of grave suffering, and the improvement of human foresight and wisdom should be pursued as urgent priorities, and heavily funded.*
6. *Policy-making ought to be guided by responsible and inclusive moral vision, taking seriously both opportunities and risks, respecting autonomy and individual rights, and showing solidarity with and concern for the interests and dignity of all people around the globe. We must also consider our moral responsibilities toward generations that will exist in the future.*
7. *We advocate the well-being of all sentience, including humans, nonhuman animals, and any future artificial intellects, modified life forms, or other intelligence to which technological and scientific advance may give rise.*
8. *We favor allowing individuals wide personal choice over how they enable their lives. This includes use of techniques that may be developed to assist memory, concentration, and mental energy; life extension therapies; reproductive choice technologies; cryonics procedures; and many other possible human modification and enhancement technologies.*

In his book Nancy [9] explores the idea of humanity as an endless denaturalization by means of science and technology. Technology is also viewed as a road from the natural toward humanization, but there is no guarantee for human survival, as humans are insufficient natural beings. This denaturalization is delineated in the Transhumanist declaration, where we have the guidelines for a new approach, capable of addressing philosophical cyborgization in all its forms: economical robotics, avatars, virtual reality, and AI singularity.

If we acknowledge some transcendental and even creationist tendencies, we can reiterate the thinking set forth in [4]. Hegelian and Marxian philosophies warn us that

[2]Original version available at https://nickbostrom.com/papers/history.pdf.

a quick and uncritical adoption of either naturalism or humanism as Nancy defined them can lead to a metaphysical dead-end. Dialectically speaking, they both embody (ontologically and anthropologically) the possibility of the unnatural and inhumane as an artificial notion which can be brought out by scientific labor. In other words—technology as means of humanization is not a technological, but a metaphysical development.

Heidegger [7] at one point considered these implications of technology as means to conquer nature. Technology does not transcend humanism, but merely deepens it.

When we talk about the creation and predestination of man, we are talking about his transhuman and posthuman essence. Behind the set, self-appropriate characters, scientific work in the name of being, human life, humanism, naturalism, mastering nature, in the name of creativity and a transhuman future, creates nature, all natural beings, including humanized man. The traditional discussion of subject–object relation ends in pseudosynthesis: subject = object, creativity = creation, freedom = alienation, naturalism = scientific naturalism, humanism = scientific humanism, human mind = superintelligence, hedonism = death of body and mind, bliss = duration on scientific technical path.

Cybernetics as the metaphysics of space—the time of scientific history, with the help of applied sciences and techniques, systematically transforms all natural beings and their interrelationships. The human being, as the medium of this creation, becomes the object itself, as the construction of a scientific and technical experiment and manipulation in order to adapt to the needs of the self-contained drive for scientific work. The existence of man and the entire cyborgized community no longer depend on the individual or the shared will of the human species, because it has become an integral part of the scientific and technical process of scientific work as a substance—a subject who, for his own ends, predetermines and uses all means and all the moments for self-reproduction.

12.3 Posthumanism and Superintelligence

Rene Descartes has painstakingly philosophized about the relationship between mind and body in his Cartesian meditations [3]. The maxim "Cogito ergo sum" establishes the essence and existence of man in the mind. The mind is a divine gift in man who shares human existence with God as his Creator. If they could separate the mind from the body, they would find the idea of the divine in it, Descartes thinks. Furthermore, Descartes thinks that the body is opposite to reason and is a disturbance to the mind in its thinking. Therefore, the essence and existence of man must be grounded in pure thought freed from imprisonment in the mortal body. Only thought without the body can become pure thought, says Descartes, reflected in geometry and arithmetic. Already here, we observe, by a later definition, a transhuman and posthuman vision of a person moving from pure thought derived from God to posthuman pure thought from a computer. Can a person exist as pure thought from a computer, or is it a life beyond and beyond man's time will show.

The absurdity of Descartes' thinking is summed up by the fact that the philosophy of pure thought of divine origin, which came to existence in a human mind, opened the way to the practice of modern anthropocentrism and materialism that would distance man from divine creation and spiritually subject him to the material. In the mid-nineteenth century, the logician George Boole set out the first formalization of laws of though after Aristotle [1]. Like Descartes, Boole believed that human thought was a bond between man and God, and that the formal description of the human mental process was at the same time a revelation of the thought of God. In this fact, Boole sees the unity of science and religion. Boolean algebra, which he thought described the most abstract and formal foundations of human thought, became the logical basis of digital computers. Thus, in a pragmatic way, Descartes' quest for pure thoughts was solved by a set of rules for the mind, leading to the idea of joining the thinking person with a thinking computer that would be analogous to the mind, but independent of it.

At that time, the idea behind a "thinking computer" was still a Cartesian mission of connecting with the immortal divine mind from a mortal human body. It was thought that with the help of a thinking computer that mimics human thinking, one would gain a lasting existence and better communication with the divine and the human. A mind free from physical limitations could develop into a higher form of artificial life and unite with its original Creator, that is, his mind (divine intelligence). The pragmatic interests of the true masters of the artificial intelligence project have shifted its development away from philosophical and theological visions and ideals. The military-industrial complex and the governments have always seen the artificial intelligence project as a quest for a practical strategic advantage.

The British logician and father of computing, Alan Turing conceived an intelligent computer as an imitator [11] which should imitate of the workings of the human mind so closely to be indistinguishable from a real human, and thereby sidelining Descartes' doubts about the importance of the natural human brain.

In the ideal case, an intelligent computer should learn, develop, augment and transcend the human mind and in time possibly replace all the faculties of the human mind. These ideas are clearly a deep inspiration for the quest for superintelligence, which has chanted its siren call to many a scientist.

There is artificial intelligence today that is more intelligent than any human in specific areas. But there are great efforts to develop a more universal intelligence—a superintelligence capable to learn, to think strategically, to deduce, and to replicate itself. Bostrom describes Superintelligence as [2] "any intelligence that far outweighs the cognitive performance of people in almost every area of interest."

Bostrom [2, p. 40] pinpoints that human level natural language understanding as the main component of superintelligence, since such an intelligence is only a few steps away from knowing everything on the Internet, and by proxy most of the knowledge of humanity as a whole. It can be argued that this is the only faculty needed for a superintelligence to arise, and Bostrom predicts that human level natural language understanding will be obtained by 2050.

There are several paths to the realization of the Superintelligence project: machine intelligence development, whole-brain emulation, biological brain enhancement by genetic or technical means, brain–computer interface development, digital network development, and many others. Which of these pathways will be the first to surpass biology in general intelligence and create strategic advantage, it is difficult to predict.

12.4 Conclusion

One might wonder how a "new world order" under the command of an "all-seeing" superintelligence might look like. This will certainly bring its own share of problems, from technical to social. Bostrom reports the estimate [2, pp. 148–149] that humans have claimed only 24% of the primary production capabilities of the world. Even at these levels, issues regarding sustainability are abundant. How would a superintelligence proceed on this front—or in numerous other systems with complex feedback loops? Would it be able not just to learn nonlinear separations, but also to predict nonlinear (and even non-continuous) phenomena? Would it discover new ways to evolve itself? Would it take the human out of the loop or consider humans to be helpful with "creativity-intensive" tasks in the same way we consider computers helpful for "calculation intensive tasks"? But all the ways the future might unfold, there is a distinct Marxian component to be observed: ontologically, superintelligence is just a natural and humane continuation of scientific work, and should it develop, it would literally represent the manifestation and realization of Hegelian and Marxian metaphysics in the world.

References

1. Boole G (1958) An investigation on the laws of thought. Dover Publications, New York
2. Bostrom N (2016) Superintelligence: paths, dangers, strategies. Oxford University Press, Oxford
3. Descartes R (2017) Meditations on first philosophy. In: Cottingham J (ed) Descartes: meditations on first philosophy: with selections from the objections and replies (Cambridge texts in the history of philosophy). Cambridge University Press
4. Greguric I (2018) Kibernetička bića u doba znanstvenog humanizma: prolegomena za kiborgoetiku. HFD, Zagreb
5. Hegel GWF (2010) Encyclopedia of the philosophical sciences in basic outline. Cambridge University Press, Cambridge
6. Hegel GWF (2010) The science of logic. Cambridge University Press, Cambridge
7. Heidegger M (2013) The question concerning technology, and other essays. Harper Perennial Modern Classics, New York
8. Marx K, Engels F (2009) The economic and philosophic manuscripts of 1844 and the Communist manifesto. Prometheus, Boston
9. Nancy J-L (2007) The creation of the world or globalization. SUNY Press, New York
10. Sutlic V (1994) Uvod u povijesno mišljenje. Demetra, Zagreb
11. Turing AM (1950) Computing machinery and intelligence. Mind 59(236):433–460
12. Wiener N (1948) Cybernetics: on control and communication in the animal and the machine. The MIT Press, Cambridge, MA

Index

S. Skansi (ed.), *Guide to Deep Learning Basics*,
https://doi.org/10.1007/978-3-030-37591-1

N

O

P

R

S

T

U

W

Printed in the United States